符号中国 SIGNS OF CHINA

中国茶

CHINESE TEA

"符号中国"编写组 ◎ 编著

中央民族大学出版社
China Minzu University Press

图书在版编目(CIP)数据

中国茶：汉文、英文 / "符号中国"编写组编著. —北京：
中央民族大学出版社, 2024.3
（符号中国）
ISBN 978-7-5660-2302-5

Ⅰ.①中⋯　Ⅱ.①符⋯　Ⅲ.①茶文化—介绍—中国—汉、英　Ⅳ.①TS971.21

中国国家版本馆CIP数据核字（2024）第016595号

符号中国：中国茶　CHINESE TEA

编　　　著	"符号中国"编写组
策划编辑	沙　平
责任编辑	李苏幸
英文指导	李瑞清
英文编辑	邱　械
美术编辑	曹　娜　郑亚超　洪　涛
出版发行	中央民族大学出版社
	北京市海淀区中关村南大街27号　邮编：100081
	电话：（010）68472815（发行部）　传真：（010）68933757（发行部）
	（010）68932218（总编室）　　　　（010）68932447（办公室）
经销者	全国各地新华书店
印刷厂	北京兴星伟业印刷有限公司
开　　本	787 mm×1092 mm　1/16　印张：12
字　　数	150千字
版　　次	2024年3月第1版　2024年3月第1次印刷
书　　号	ISBN 978-7-5660-2302-5
定　　价	58.00元

版权所有　侵权必究

"符号中国"丛书编委会

唐兰东　巴哈提　杨国华　孟靖朝　赵秀琴

本册编写者

羽　叶

前言 Preface

　　茶为中国国饮，中国是茶的故乡。中国人最早发现、栽培和饮用茶叶，并将茶树种子、采制技术、品茶技艺等传播到全世界。茶兴于中国的唐代，在丝绸、瓷器向

　　Tea is the national drink of China. China is the hometown of tea. Chinese people were the first to discover, cultivate, and drink tea. They spread tea seed, tea picking and processing techniques, and tea appreciation techniques to the rest of the world. Tea began to gain fame in the Tang Dynasty (618-907) and was gradually exported to other countries together with other specialties of China such as silk and chinaware. It was loved by more and more people in the world. Today, with people paying greater attention to health, Chinese tea is popular again due to its unique health-care efficacy.

　　There are numerous tea varieties in China and many of them are famous ones. In ancient China, people associated tea drinking with

国外输出的同时，茶叶也逐渐传播至国外，并越来越受到外国人的青睐。时至今日，随着人们对健康的重视，中国茶更以其独特的保健功效再一次掀起了饮茶热潮。

中国茶种类繁多、名茶荟萃。在中国古代，人们将饮茶与品行修养联系在一起，讲求品茶悟道的文化情韵。而以此为基础诞生的茶文化博大精深，茶人茶事、茶礼茶俗、茶诗茶画，更是包罗万象。承载着几千年历史的中国茶文化，正迈着轻盈的步履将万缕茶香向全世界散播。

本书以中英文对照的形式，配以大量精美图片，向海内外读者展示了中国茶的历史、茶中名品、泡茶的方法以及一些茶艺茶俗，便于读者了解中国茶、感受中国茶文化的那一份悠长。

character building and laid stress on the cultural charm in tea appreciation and philosophical realization. The tea culture against such background has been extensive and profound. It involves all the things related to tea, including people and their tea-related anecdotes, tea protocols and custom, and tea poems and paintings. Carrying the several-millennium Chinese history, the tea culture of China is being spread to the whole world with the penetrating aroma of the tea.

Containing large quantities of exquisite pictures and bilingual illustrations (Chinese and English), this book presents to readers at home and abroad the history of Chinese teas, famous tea varieties, the way of tea making, and some tea ceremonies and custom. Through it, the readers can better understand Chinese teas and the time-honored tea culture of China.

目 录 Contents

茶的故乡
Hometown of Tea 001

最早的茶树
The Earliest Tea Tree 002

饮茶的历史
History of Tea Drinking 011

陆羽与《茶经》
Lu Yu and *The Classic of Tea* 016

中国人爱饮茶
Chinese People's Love for Drinking Tea 029

名山出好茶
Good Teas from Great Mountains 037

武林山龙井茶
Longjing (Dragon Well) Tea from Mt. Wulin 038

顾渚山紫笋茶
Zisun (Purple Bamboo Shoot) Tea
from Mt. Guzhu 047

普陀山佛茶
Buddha Tea from Mt. Putuo 050

洞庭山碧螺春
Biluochun (Green Spiral Spring) Tea
from Mt. Dongting.. 053

黄山毛峰茶
Maofeng (Hairy Tip) Tea
from Mt. Huangshan 057

庐山云雾茶
Yunwu (Cloud and Mist) Tea
from Mt. Lushan.. 061

峨眉山竹叶青茶
Zhuyeqing (Green Bamboo Leaf) Tea
from Mt. Emei 065

蒙山蒙顶茶
Mengding (Mt. Mengshan Summit) Tea
from Mt. Mengshan 069

君山银针茶
Yinzhen (Silver Tip) Tea from Mt. Junshan 074

六大茶山普洱茶
Pu'er Tea from Six Major Tea Mountains 077

太姥山绿雪芽茶
Lvxueya (Green Snow Bud) Tea
from Mt. Taimu.. 084

武夷山大红袍
Dahongpao (Big Red Robe) Tea
from Mt. Wuyi 090

凤凰山单丛茶
Dancong (Single Clump) Tea
from Mt. Phoenix.. 097

冻顶山乌龙茶
Oolong Tea from Mt. Dongding........................ 101

名水泡名茶
Famous Water Making Famous Tea 103

庐山谷帘泉
Gulian (Grain Curtain) Spring in Mt. Lushan 104

济南趵突泉
Baotu (Spouting) Spring in Jinan City 108

崂山矿泉
Mineral Spring in Mt. Laoshan 112

无锡惠山泉
Huishan (Favorable Mountain)
Spring in Wuxi City 115

镇江中泠泉
Zhongling (Middle Cold)
Spring in Zhenjiang City 118

虎丘第三泉
The Third Spring in Mt. Huqiu 121

峨眉玉液泉
Yuye (Jade Liquor) Spring in Mt. Emei 123

杭州虎跑泉
Hupao (The Dreamed Tiger Pawed)
Spring in Hangzhou City 125

杭州龙井泉
Longjing (Dragon Well)
Spring in Hangzhou City 127

长兴金沙泉
Jinsha (Gold Sand) Spring in
Changxing County 129

名器配名茶
Famous Tea Set for Famous Tea..................... 131

陶质茶具
Ceramic Tea Set 132

瓷器茶具
Porcelain Tea Set 134

金属茶具
Metallic Tea Set 139

漆器茶具
Lacquer Tea Set 141

竹木茶具
Bamboo-Wood Tea Set 143

如何泡好茶
How to Make a Good Cup of Tea 149

冲泡绿茶
Brewing Green Tea 150

冲泡白茶
Brewing White Tea 156

冲泡黄茶
Brewing Yellow Tea 160

冲泡乌龙茶
Brewing Oolong Tea 162

冲泡红茶
Brewing Black Tea 166

冲泡黑茶
Brewing Dark Tea 169

茶的故乡
Hometown of Tea

世界各国最初所饮的茶叶，以及引种的茶种、饮茶的方法、栽培的技术等都是直接或间接地从中国传播出去的。在中国发现的野生大茶树，时间之早、树体之大、数量之多、分布之广，可谓世界之最。最早的"茶"字出自中国，世界各国有关茶的读音，也都源于中国。中国是茶的发祥地。

All countries in the world received their first import of tea leaves, tea seeds, tea drinking methods, and tea cultivation techniques directly or indirectly from China. The wild large tea trees found in China rank first in the world in terms of age, size, number, and distribution scope. The word "tea" is from China. The pronunciation for tea in many languages are found on Chinese character 茶 (tea). China is the birthplace of tea.

> 最早的茶树

1824年，驻印度的英国少校勃鲁士（R.Bruce）在印度阿萨姆省沙地耶（Sadiya）发现了一株高约43英尺（约13.1米）、直径3英尺（约0.9米）的野生茶树。为此，西方学者认为印度是茶的故乡。

> The Earliest Tea Tree

In 1824, British explorer Major Robert Bruce discovered a wild tea tree some 43 feet (about 13.1 meters) tall and three feet (about 0.9 meters) in diameter in Sadiya of the Assam region in India. Therefore, western scholars deduced that India was the hometown of tea.

In fact, *The Classic of Tea* recorded a tea tree about one meter in diameter in the Tang Dynasty(618-907). About the same size as the one found by Rober Bruce, but some 1,100 years earlier.

In 1961, Chinese archeologists discovered in Yunnan Province a wild

- 千年古茶树
  The Millennium Ancient Tea Tree

神农氏与茶

神农氏是中国古代传说中的神话人物。中国的先民认为他不但是农业和医药的发明者,也是茶的发现者,因此把中国的农耕生产、饮食和医药文化的起源都归功于他。

相传神农氏有一个透明的肚子,可以看见自己的五脏六腑。他为世人尝遍百草,凡是吃到肚中的食物,都看得一清二楚,能分辨出哪些植物有毒,哪些植物可以食用,哪些植物可以当做药用。一天,神农氏尝了七十二种毒草,他的五脏发黑,很快就要支撑不住了。就在这时,忽然看见身旁有一丛灌木,翠绿的叶子带着幽幽的清香,于是他采下几片放入口中。很快,叶汁在他的肚子里流动,五脏马上恢复原样。这丛灌木就是茶树。神农氏认识到茶叶的解毒功能,于是将其推广给世人作为药用。

Shennong and Tea

Shennong was a legendary figure in ancient Chinese myth. The ancient Chinese regarded him as not only the inventor of agriculture and medicine, but also the finder of tea. They therefore attributed the origin of China's agricultural production, diet, and medical culture to him.

According to the legend, Shennong had a crystal belly, so that everything went on in the stomach could be clearly seen from the outside. He tasted all kinds of plants, in order to let people know which things can be eaten, which are poisonous and which can help cure disease. One day, Shennong was poisoned by 72 different poisonous plants. His internal organs turned black and he was dying. Then he saw a clump of bush with green scented leaves. He picked several leaves and put them into his mouth. With the leaf juice ran through his stomach, his internal organs returned to normal at once. Those leaves were from tea trees. Thus Shennong discovered the detox function of tea leaves and recommended it to people as an antidote to poisons.

• 《神农百草图》【局部】

图中神农身背药篓,左手拿采药工具,右手执药草,腰围树叶,赤足而行,展现出神农采百草的情景。

The Painting of Shennong and Hundreds Herbs [Part]

In the picture, Shennong carries a basket on his back. He holds a picker in his left hand and a herb in his right hand. With tree leaves tied around his waist, he walks barefooted on an herb-collecting trip.

而事实上，中国唐代《茶经》中就曾记载过当时中国已发现有直径约1米的茶树，大小与勃鲁士发现的那棵相当，而时间上却要早了大约1100年。

1961年，中国考古学家在云南省发现了一棵高达32.12米的野生大茶树，树龄达1700年左右，其树高和树龄在山茶属植物中均属世界第一，是目前已发现最大、最古老的野生大茶树。

茶树属山茶属植物，山茶属是比较原始的植物种群，而茶树又是山茶属中一个比较原始的种，其起源据推测至少在6000万到7000万年前。世界上山茶属植物共有23属，380余种，而在中国就有15属，260余种。

中国是野生大茶树发现最早、最多的国家。云南、贵州和四川地区的茶树较多。这些地区的茶树多属高大乔木树型，具有较典型的原始形态特征，说明中国的西南地区是山茶属植物的发源中心，是茶的发源地。

large tea tree which was 32.12 meters tall and about 1,700 years old. Its height and age both were world No.1 among camellia plants. It still remains the largest and oldest wild tea tree that has been found in the world.

The tea tree is of the species Camellia sinensis. Camellia is a rather primitive plant species and the tea tree is a relatively primitive member of the genus Camellia. According to research, the tea plant evolved at least 60-70 million years ago. There are 23 genera and over 380 varieties of camellia plants in the world, of which, 15 genera and over 260 varieties are found in China.

China is the country which first discovered wild big tea trees and has the most numerous wild large tea trees. Tea trees in China are normally found in Yunnan Province, Guizhou Province and Sichuan Province. As large and high arbors in these areas, the tea trees show typical primitive features, indicating that the southwestern area of China is the birthplace of Camellia plants, including tea.

- 云南风光

云南、贵州和四川地区是茶树的原产地。这里地形复杂，大部分为盆地、高原地势。雨量充沛、冬暖夏凉的气候以及肥沃的土壤为茶树的生长提供了条件。

Landscape of Yunnan Province

Yunnan Province, Guizhou Province and Sichuan Province are the places of origin of tea trees. These areas have a complicated terrain, which mainly includes basins and plateaus, and enjoy an abundant rainfall, a climate warm in winter and cool in summer and fertile soil, all being favorable for the growth of tea trees.

茶树

　　茶树是多年生常绿木本植物，生长于中国长江流域各省，后传入日本、印度尼西亚、斯里兰卡、俄罗斯等国。茶树喜温、喜湿、喜阴，适合生长在气候湿润、雨水充沛、多云雾、少日照、土壤肥沃的地方。

　　茶树分为乔木、半乔木、灌木三种类型。乔木型茶树植株高大，通常高3～5米，主干直立明显，分枝部位较高，主根发达；半乔木型茶树植株高度中等，分枝部位离地面较近，主干较明显，根系较发达；灌木型茶树植株矮小，通常只有1.5～3米，近地面处枝干丛生，或从根茎处发出，分枝稠密，成年后无明显的主干，根系分布较浅，侧根发达。

Tea Tree

The tea tree is a perennial evergreen woody plant that originates in the provinces of China along the Yangtze River. It has been introduced to other countries such as Japan, Indonesia, Sri Lanka, and Russia. The tea trees like warm, humid and shady environment. It thrives in the regions with humid climate, abundant rainfall, cloudy and misty weather, less sunshine, and fertile soil.

　　There are three types of tea trees, namely arbor-type tea tree, semi-arbor-type tea tree, and shrub-type tea tree. The arbor-type tea tree is normally large and tall, as tall as three to five meters. Its main trunk is obviously upright and its branches grow from a relatively high position. It has a highly developed root system. The semi-arbor-type tea tree has a medium height. Its main trunk is rather obvious and its branches grow from a position rather close to the ground. It has a relatively developed root system. The shrub-type tea tree is small and short. It is only 1.5 to 3 meters tall. Its branches grow thickly near the ground or out of the stem. The mature tea tree has no obvious main trunk. Its root system is distributed rather shallowly and its side roots are highly developed.

• 茶树
A Tea Tree

- 乔木型茶树
  An Arbor-type Tea Tree

- 灌木型茶树
  Shrub-type Tea Trees

- 半乔木型茶树
  A Semi-arbor-type Tea Tree

茶马古道

茶马古道起源于唐宋时期的"茶马互市",即以茶易马或以马换茶。这是汉藏民族之间一种传统的贸易往来。茶马古道是因茶与马的交换而形成的交通要道。历史上的茶马古道不止一条,它以川藏道、滇藏道和青藏道三条大道为主线,辅以众多的支线,构成一个庞大的交通网络。

Ancient Tea-horse Road

The Ancient Tea-horse Road stemmed from the traditional Tea-Horse Trade by Barter between Han and Tibetan peoples in the Tang Dynasty(618-907) and the Song Dynasty(960-1279). That were mainly tea-for-horse or horse-for-tea trades at that time. There were more than one ancient Tea-horse Road in history. In addition to three major roads, namely Sichuan-Xizang Road, Yunnan-Xizang Road, and Qinghai-Xizang Road, there were numerous smaller branch roads. They formed a massive transportation network.

- 行走在茶马古道上的马帮
  A Caravan on the Ancient Tea-horse Road

"茶"字及其发音

"茶"字是由"荼"字演化而来。"荼"是一种苦菜。唐代时,读音已经由"tú"转为了"chá",但是写法没有改变。有学者认为,"茶"字的最早出现大约是在唐宪宗元和年间。由于茶叶生产的发展,饮茶的人越来越多,书写"荼"字的频率也越来越高,有民间书写者就把"荼"字减去一画,成了现在的"茶"字。

- 古时茶称为"荼",是苦菜的一种
  In the Ancient Times, Tea Was Called 荼 (Pronounced as tú), a Kind of Bitter Vegetable

荼 → 茶

- From 荼（tú）to 茶（chá）

中国地域辽阔,方言多种多样,各地对茶字的读音有所不同。如在广东省,广州附近发"chá"音,到了汕头却发"tè"音。又如福建省,茶在福州的发音是"tá",到了厦门却近似汕头的"tè"音。长江流域及华北地区又有"chái""zhou""chá"等不同发音。

"茶"的读音是随着茶叶在世界各地的传播而流传海外的,所以世界各国有关茶的读音也源于中国,主要有两大体系,一是普通话"茶"音,即"chá";一是福建厦门地方语"tè"音,即"tey"。

"cha"音主要传往中国的周边国家，日本直接使用汉字"茶"，古波斯为"cha"，土耳其语为"chay"，俄语为"Чай"，蒙古语为"chai"，伊朗语为"chay"，朝鲜语为"sa"，阿拉伯语为"chay"。

明末清初，西方远洋船队在厦门沿海地区设有办事处，把当地人对茶的称呼"tè"直接译过去，便译成英语的"tee"、拉丁语的"thea"，后来英语拼成"tea"。至于法语系的"thé"，德语系的"tee"，荷兰语的"thee"，意大利语的"te"，西班牙语的"té"，希腊语的"tsai"，南印度的"tey"，斯里兰卡的"they"等，都是由厦门话"tè"音和英语的译音演变而成的。

Chinese Character 茶 (Tea) and Its Pronunciation

The Chinese character 茶 (tea, pronounced as chá) evolved from the character 荼 (pronounced as tú), a kind of bitter vegetable. In the Tang Dynasty (618-907), the Chinese pronunciation for tea had changed from "tú" to "chá", but the style of writing remained the same. Some scholars believe that the character 茶 (tea) first emerged in the Tang Dynasty during the Yuanhe Period (806-820) under the reign of Emperor Xianzong. With the development of tea production, more and more people began drinking tea. The Chinese character for tea was used more and more frequently. Some writers took one horizontal off the character 荼 and gave birth to present-day character 茶 (tea).

China is a large country with diversified dialects. The Chinese character for tea is pronounced differently in different areas. In Guangdong Province, it sounds like "chá" around Guangzhou and "tè" around Shantou. In Fujian Province, it is pronounced as "tá" in Fuzhou and again "tè" in Xiamen. In the area along the Yangtze River and in North China, it has other different pronunciations such as "chái", "zhou" and "chá".

The Chinese pronunciation for tea was spread to other countries with the export of tea. Therefore, the pronunciation of tea in the world also originated from China. There are two major systems. One is mandarin pronunciation as "chá" and the other Xiamen dialect pronunciation in Fujian as "tè" or "tey".

"Cha" the Chinese pronunciation for tea, was mainly spread to the countries around China. In Japan, the Chinese character 茶 (tea) is directly used. Tea is called "cha" in ancient Persian, "chay" in Turkish, "Чай" in Russian, "chai" in Mongolian, "chay" in Iranian, "sa" in Korean, and "chay" in Arabic.

In late Ming Dynasty and the early Qing Dynasty, western fleets set up offices in Chinese coastal areas near Xiamen. They adopted "tè", local pronunciation for tea, and invented a new pronunciation of "tee" in English and "thea" in Latin. Later, the English word "tea" came into being. In fact, "thé" in French, "tee" in German, "thee" in Dutch, "te" in Italian, "té" in Spanish, "tsai" in Greek, "tey" in the language used in the south of India, and "they" in Sri Lankan all evolved from the pronunciation "tè" in Xiamen dialect and the English pronunciation of tea.

> 饮茶的历史

在神农氏之后，人们发现茶不仅有解毒的功能，还有助于消化，于是，就把茶与其他食物一起加工，当做菜吃。再后来，随着社会的进步与发展，人类生活得到改善，茶叶被加工后烹煮饮用，成为饮品。

中国史籍上有"茶兴于唐"的说法，唐代被认为是茶的黄金时代。

唐代以前，南方盛行饮茶，而北方人却厌恶茶，认为喝茶是一种奇风怪俗，甚至以之为耻。而到了唐代，贡茶推动了茶的兴盛。著名的贡茶有产于浙江长兴的顾渚紫笋茶和产于江苏宜兴的阳羡茶。朝廷将制茶中心从巴蜀地区转移到江南地区，促进了江南制茶技术的提

> History of Tea Drinking

In addition to tea's detoxifying effect introduced by Shennong, people found out that tea also helped digestion. They therefore processed tea together with other food and served as dishes. Later, with social progress and development, people enjoyed a better life and tea was processed and cooked into beverage.

According to Chinese historical records, tea began to gain fame in the Tang Dynasty (618-907). The Tang Dynasty is regarded as the golden era for tea.

Before the Tang Dynasty (618-907), people in South China loved drinking tea while people in the north hated tea and regarded tea drinking as a strange custom or even a shameful practice. By the Tang Dynasty, tribute teas promoted the rise of tea. The famous tribute teas included the Zisun Tea produced in Mt. Guzhu of Changxing County, Zhejiang Province and

● 《清明茶宴图》（唐）
唐代时宫廷兴办清明茶宴。在清明前后皇帝收到贡茶后，要先祭祀祖宗，后赐给近臣宠侍，并摆"清明茶宴"以飨群臣。

The Painting of a Tea Banquet at the Qingming Festival (Tang Dynasty 618-907)

In the Tang Dynasty, the imperial court often held tea banquet at the Qingming Festival (Pure Brightness/Tomb-Sweeping). Every year in early April, emperors received tribute teas from various places. First, some would be offered as a sacrifice to their ancestors. Then, some would be given to their favourite officials and servants as rewards. And held Royal Qingming Tea Banquets to entertain high rank officials.

高，带动了全国茶叶的生产与发展。从此，饮茶之风遍及大江南北，塞内塞外。

唐代茶分为粗茶、散茶、末茶、饼茶，以饼茶为主。品饮方式以煮茶法为主，即先将饼茶烤干以蒸发其中的水分，干后装袋以保持茶香，待饼茶冷却后将其碾成细末待煮。煮茶时会加入盐，有的还添加其他香料。

the Yangxian Tea produced in Yixing City, Jiangsu Province. The imperial government moved the tea manufacturing center from Sichuan Province to the area south of the Yangtze River, which promoted the tea-making techniques in the new center and brought along the tea production and development in the whole country. Since then, tea drinking has become popular in the whole country.

In the Tang Dynasty (618-907), tea

宋代，民间饮茶之风盛行，并逐渐发展出以茶款客的礼仪。当时的文人雅士都喜爱喝茶、推崇饮茶，并以相聚品茗为雅，进一步推动了饮茶之风的蔓延。

was divided into raw tea, loose tea, dust tea, and caky tea. Among them, caky tea was the main type. Tang people made tea by cooking it. First, they dried the caky tea by baking. Then, they bagged the tea to keep its aroma. Later, after the tea had cooled, they ground it into fine powder for cooking, during which, they added salt or other spices.

In the Song Dynasty (960-1279), tea drinking was popular among ordinary people. Gradually, the etiquette of treating guests with tea came into being. At that time, all refined scholars liked tea and praised it highly. They regarded tea drinking as a social grace, which promoted tea drinking as a fashion.

- 《调琴啜茗图》【局部】周昉（唐）
  The Painting of Zither Playing and Tea Drinking [Part], by Zhou Fang (Tang Dynasty 618-907)

- 《撵茶图》刘松年（宋）
  The Painting of Tea Grinding, by Liu Songnian (Song Dynasty 960-1297)

宋代以末茶、散茶为主，但团茶、饼茶依然受到人们的喜爱，而且在蒸压团茶时，会加入龙脑等名贵香料，尤其是出现了皇家专用的"龙凤团茶"，表面压制有龙凤图案。饮茶方法以点茶为主，即先烤茶饼，再敲碎，碾成细末，用茶罗将茶末筛细。茶末放于碗中，倒入少量沸水调成糊状。用釜烧水，微沸初漾时，即将开水冲入杯、盏、碗内。为使茶末与水相融，需用打茶工具大力搅拌，以渐起泡沫为佳。

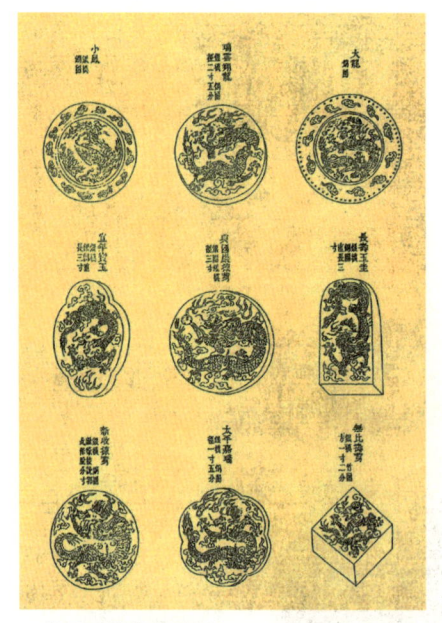

- 龙凤团饼茶线描图
A Line Drawing of the Dragon-phoenix Lump Tea

In the Song Dynasty (960-1279), dust tea and loose tea were very popular, but lump tea and caky tea were still loved by people. In producing the lump tea, precious spices such as borneol could be mixed in. There was also the dragon-phoenix lump tea exclusively made for the royal family. This kind of tea had dragon and phoenix patterns on its surface. The process of making tea was breaking the baked caky tea and then shifting the ground small with tea sieve, putting the tea powder in a bowl or cup and mixing with small amount of hot water to get a paste, boiling water in a kettle, then infusing the boiling water into the bowl or cup as soon as the water just starts boil. For fully dissolving the tea powder in the water, a tea-beater was used to stir hard until foam emerging.

In the Ming and Qing dynasties, caky tea was gradually abolished as a tribute. Instead, loose tea production was popular. Its making and drinking methods were quite different from the previous one. Boiling was replaced by brewing. A pinch of tea leaves was put into a tea cup or teapot and boiling water was poured in to make tea. At that time, there were many brewing methods. Putting the tea in first then pouring water on top was

明清时期逐渐废除了饼茶进贡，流行炒制散茶，其冲饮方法也与此前有了很大的不同，改煮为泡，直接将一小撮茶叶放入茶杯或茶壶，倒入开水即成。当时茶有多种泡法，先放茶后注水为"下投"；放一半水再放茶再注满水为"中投"；先注满水再放茶为"上投"。

明清时期散茶的崛起促进了各地名茶的出现，也为茶艺的出现提供了有利的条件。泡茶讲究用水和茶器，茶、茶具、茶艺等在这一时期得到了全面的发展。

到了现代，中国人饮茶的主要方式是清饮法，即以开水直接冲泡茶叶的方法。除此之外，也出现了一些新的内容和形式，如调饮法、袋泡茶、听装茶等。

called "Bottom Throw" (*Xiatou*). Putting in a small amount of water followed by tea then more water was called "Middle Throw" (*Zhongtou*), Putting the leaves on top of the water was called "Top Throw" (*Shangtou*).

The rise of loose tea in the Ming and Qing dynasties led to the emergence of the famous local teas and offered favorable conditions for the advent of tea art. Tea brewing relied heavily on water and tea wares. Tea, tea set, and tea art had an all-around development during this period.

In modern times, Chinese people drink tea by brewing tea leaves directly in boiling water. There are also other new contents and forms of tea drinking, including tea flavoring method, bag tea, and canned tea.

- 描金粉彩饰家族纹章装茶具（清）
  Powdered-color Tea Set Bearing Family Impress with Patterns Painted in Gold (Qing Dynasty 1616-1911)

> 陆羽与《茶经》

《茶经》是世界上第一部关于茶的著作,是由中国茶道的奠基人陆羽所著。《茶经》成书于公元750年前后,系统地介绍了唐代及唐代以前的茶叶历史、产地、功效、栽培、采制、煎煮、饮用等知识。

- 《茶经》
  The Classic of Tea

> Lu Yu and *The Classic of Tea*

The Classic of Tea is the first book on tea in the world. It was written by Lu Yu, founder of Chinese teaism. Completed around 750 A.D., it systematically introduced history, producing origins, efficacy, cultivation, harvesting, processing, brewing/infusing methods and drinking of tea in and before the Tang Dynasty (618-907).

Lu Yu, also known by his style name Hongjian, was a native of Jingling of Fuzhou (present-day Tianmen City of Hubei Province) lived in the Tang Dynasty.

According to the legend, Lu Yu was an orphan abandoned by his parents. He was adopted by a Buddhist monk, Zhiji, abbot of the Longgai Temple in Jingling. A smart boy by nature, Lu Yu learnt how to prepare tea for Zhiji at 8 years old. When he was 12, he fled the temple

陆羽，字鸿渐，唐代复州竟陵（今湖北天门）人。

相传陆羽是个被遗弃的孤儿，是竟陵龙盖寺的住持智积禅师收养了他。陆羽天资聪颖，8岁时就学会了煮茶，侍奉智积禅师。在他12岁时，因不愿意削发为僧，从龙盖寺逃离出来，到了一个戏班子学演戏，作了优伶。他虽其貌不扬，又有些口吃，却幽默机智，演丑角极为成功，还编写了三卷笑话书《谑谈》。后来，陆羽出众的表演才能受到竟陵太守李齐物的欣赏。李齐 because he was unwilling to become a monk. A theatrical troupe took him in and taught him acting. Though ugly and speaking with a little stutter, he was a successful clown due to his fine sense of humor and resourcefulness. He even wrote three volumes of farces entitled *Talk of Jests*. Later, his outstanding performance drew the attention of Li Qiwu, the mayor of Jingling, who recommended him to Mr.Zou, a learned hermit in Mr.Huomen.

Lu Yu made some achievements in literature and had great interest in geography and tea. He traveled everywhere in Hubei Province and Sichuan Province for investigating local geography, tea, and water. To avoid the chaos caused by war, he went to Shengzhou (present-day Nanjing City of Jiangsu Province) and lived in the Qixia Temple, where he delved in perfecting his skill in tea. After leaving the Qixia Temple, he lived as a hermit in Shaoxi(present-day Wuxing City of

- 茶圣陆羽
  Lu Yu, the Sage of Tea

物修书一封推荐他到隐居于火门山的邹夫子那里学习。

后来，陆羽在文学上小有成就，又开始对地理和茶叶产生了浓厚的兴趣。于是他出游巴山峡川。一路上风餐露宿，考察地理，评茶论水。不久之后为避战乱，陆羽来到升州（今江苏南京），寄居栖霞寺，一心钻研茶事；后又隐居苕溪（今浙江吴兴），与得道名僧皎然结为忘年之交。在皎然的帮助下，他全力展开对吴兴人文、历史、地理、茶叶的考察研究，几近痴迷，最终写成传世著作《茶经》，并终老于茶乡湖州。

陆羽逝世后，人们尊称他为"茶圣"。

• 《陆羽烹茶图》【局部】赵原（元）
The Painting of Lu Yu Cooking the Tea[Part], by Zhao Yuan (Yuan Dynasty 1206-1368)

Zhejiang Province) and made friends with the accomplished monk Jiao Ran despite their great age difference. With the help of Jiao Ran, he wholeheartedly carried out investigation and study of the people, history, geography and tea of Wuxing. Finally, he completed his great book *The Classic of Tea*. He died in Huzhou, a famous hometown of tea.

After his death, Lu Yu was praised as the Sage of Tea.

Before *The Classic of Tea* came out, tea was normally cooked together with rice and other food. It was seldom consumed as a separate drink. Lu Yu believed that such practice concealed tea's natural aroma. He thus recommended direct tea drinking in his *The Classsic of Tea*, which has since remained as the commonest tea drinking method. The advent of *The Classic of Tea* promoted the popularity of tea drinking and upgraded the art of tea drinking.

The Classic of Tea has ten chapters. They are *Chapter One: The Origin of Tea, Chapter Two: Tools of Tea-Production, Chapter Three: Process of Producing Tea, Chapter Four: Tea Wares, Chapter Five: Tea Brewing/Infusing Methods, Chapter Six: Ways of Tea-Drinking, Chapter Seven: Anecdotes*

在《茶经》问世之前，茶通常是与饭食同煮的，很少单独饮用。陆羽认为这样的煮法掩盖了茶叶原有的清香味道，于是在《茶经》中提出直接清饮的饮茶方法。这一饮茶方法成为后来最常用的饮茶方法。《茶经》的问世，推动了饮茶的风气和饮茶艺术的提高。

《茶经》共分为十章，分别是《一之源》《二之具》《三之造》《四之器》《五之煮》《六之饮》《七之事》《八之出》《九之略》《十之图》。

of Tea, Chapter Eight: Tea Producing Regions, Chapter Nine: Omissions of Tea Making, Chapter Ten: Illustrauions of Tea.

Chapter One: The Origin of Tea introduces the origin, forms, names, quality, and efficacy of tea.

Chapter Two: Tools of Tea-production introduces the tools used for picking and producing tea and their usage.

Chapter Three: Process of Producing Tea introduces the methods of picking, processing, classifying and differentiating for the steamed green tea.

Chapter Four: Tea Wares expounds in detail on the names, shapes, structures, sizes, producing methods and purposes of the tea wares for brewing and drinking tea, and their impact on tea. It also discusses the quality of the tea sets in different places and their use rules.

- 《茶经》
The Classic of Tea

《一之源》介绍了茶的起源、性状、名称、品质、功效等。

《二之具》介绍了采茶和制茶用的工具及使用方法。

《三之造》介绍了蒸青绿茶的采制、加工、分类及鉴别方法。

- 茶花图
  The Painting of Tea Tree Flowers

《四之器》详细叙述了煮茶和饮茶用具的名称、形状、用材、规格、制作方法、用途，以及器具对茶汤品质的影响等，还论述了各地茶具的好坏及使用规则。

《五之煮》重点介绍了烤茶的方法、泡茶用水和煮茶火候，以及煮沸程度和方法对茶汤色、香、味的影响。

《六之饮》重点介绍了从采摘到饮用的整个过程以及注意事项。

《七之事》以人物为线索，介绍了从神农氏到徐绩，跨越3000余年的

Chapter Five: Tea Brewing/Infusing Methods introduces the tea baking method, ideal water for tea brewing and proper heat condition for tea boiling. As well as different phases of water boiling and methods affect the color, aroma, and taste of the tea.

Chapter Six: Ways of Tea-Drinking introduces the entire process from tea leaves picking to tea drinking and the points for attention.

Chapter Seven: Anecdotes of Tea introduces the tea-related affairs during the 3,000-year period about some historical figures from Shennong to Xu Ji. By quoting the tea-related historical

- 《煮茶图》【局部】王问（明）
  图中描绘的是用竹炉煮茶的场景。
  The Painting of Tea Boiling [Part], by Wang Wen (Ming Dynasty 1368-1644)
  The painting depicts the scene of tea boiling with a bamboo stove.

• 《品茶图》陈洪绶（明）
The Painting of Tea Appreciation, by Chen Hongshou (Ming Dynasty 1368-1644)

茶事，记述了唐代以前与茶相关的历史资料，包括传说、典故、诗词、杂文、药方等，勾勒出一幅有关唐代以前中国社会饮茶风情的画卷。

《八之出》，陆羽根据自己亲身考察的经历，对各地茶叶的优劣进行评定。他将唐代全国茶区分为八大块，分别是山南、淮南、浙西、剑南、浙东、黔中、江南以及岭南，并以上、中、下、又下四个级别，对每一茶区不同地方所产茶叶质地进行评定。

《九之略》从不同条件、时间一一介绍茶具、茶器和制茶、煮茶的省略情况。

data before the Tang Dynasty (618-907), including legends, allusions, poems and verses, essays and prescriptions. It gives the profile of tea drinking custom in Chinese society before the Tang Dynasty.

Chapter Eight: *Tea Producing Regions* contains Lu Yu's appraisal of teas from different regions based on his personal investigation. According to his division, there were eight major tea-producing regions in the Tang Dynasty (618-907): Shannan (the area south of Mt. Gangdisi and Mt. Nianqingtanggula), Huainan (the area south of the Huaihe River), Zhexi (the area west of Zhejiang Province), Jiannan (the area northwest of Chengdu Plain), Zhedong (the area east of Zhejiang Province), Qianzhong (the center of Guizhou Province), Jiangnan (the area south of the Yangtze River), and Lingnan (the area south of China's five southern mountains). Lu Yu graded teas from these regions into four levels, namely top, medium, low, and lowest.

Chapter Nine: *Omissions of Tea Making* introduces some tea sets, tea wares, tea making steps can be omitted according to different conditions and times.

Chapter Ten: *Illustrations of Tea* teaches people copy the above nine

《十之图》是用白色绢子四幅或六幅，分别把以上九章写在上面，张挂于墙壁之上。这样，对茶的起源、制茶工具、茶的采制、烹饮茶具、煮茶方法、茶的饮用、历代茶事、茶叶产地、茶具使用，都会看在眼里，牢记于心，对《茶经》的内容可一目了然。

chapters on four or six pieces of white silk and put them on the wall. Therefore, the contents of *The Classic of Tea* would be clear at a glance. It would be easier to remember the origin of tea, tea-producing tools, tea picking and processing, tea wares, tea-making methods, tea drinking, anecdotes of tea, tea producing regions and omissions of tea sets.

《茶经》中茶叶的加工工序

　　蒸熟：在加工前，先要把新鲜的茶叶放在水里浸泡、洗涤，然后再放在蒸笼里蒸，这是为了去除叶片上沾染的灰尘，同时降低茶叶的苦涩感。

　　捣碎：蒸好的茶叶在捣碎前，还得先压榨去汁，以确保茶味不涩，才能放入瓦盆内捣烂、研细。

　　入模拍压成形：模子形状不一，故团茶的形状有多种。拍压需在石制的承台上进行，先将襜布放在承台上，将茶膏倒入模中，将模子放在襜布上，不断拍击，使其结构紧密坚实。成形后，拉起襜布取出茶团。

　　焙干：拍压成形后的团茶要马上进行烘焙，以防变质。

　　穿成串：用锥刀、竹条把烘焙后的团茶串在一起。

　　封装：制好的团茶必须及时、正确封装储存。封装好的茶叶一般放在育器中。育器以木头为框架，四周用竹篾编成竹壁，竹壁用纸裱糊，中间设有火盆，盛有炭火灰，可保持干燥。江南梅雨季节时用此烧明火去湿。

Tea Processing Procedures in *The Classic of Tea*

Steaming: Before processing, the fresh tea leaves should be soaked in water and rinsed. Then, they are steamed in a steamer to get rid of the dust and reduce their harsh taste.

　　Smashing: The steamed tea leaves should be pressed to get rid of its juice and the astringent taste. Then, they are smashed in a ceramic basin.

　　Mould beating and pressing: Different shapes of moulds turn out different shapes of lump tea. Pressing should be done on a stone platform. Put an apron on the platform, pour the tea paste into the mould, put the mould on the apron, and beat it continuously to make it compact and solid. After the formation, pull up the apron and take out the tea lump.

Baking: The lump tea should be baked immediately after formation against deterioration.

Stringing up: The baked lump tea is stringed up with an awl and bamboo strips.

Packing: The finished lump tea should be packed for storage timely and correctly. The packed tea is normally placed in a container called Yu, which has a wooden framework, a bamboo-splint wall and covered with paper, and a brazier containing burnt charcoal ash in the center to keep the tea dry. During the plum rain season in the area south of the Yangtze River, people use open fire to get rid of dampness.

唐代饼茶制作流程
Procedures for Making Caky Tea in the Tang Dynasty (618-907)

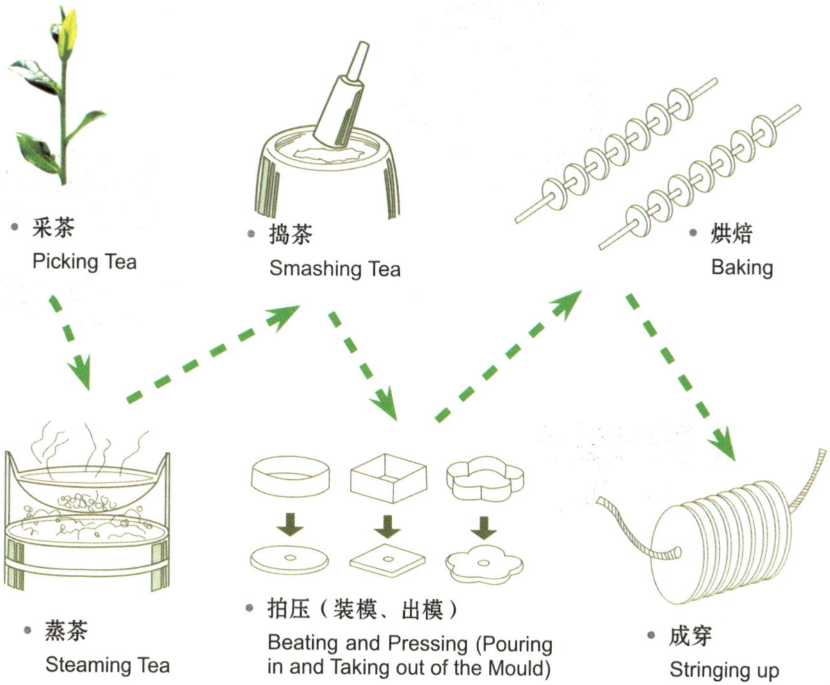

- 采茶 Picking Tea
- 捣茶 Smashing Tea
- 烘焙 Baking
- 蒸茶 Steaming Tea
- 拍压（装模、出模）Beating and Pressing (Pouring in and Taking out of the Mould)
- 成穿 Stringing up

清代茶叶制作过程
Tea Processing Procedures in the Qing Dynasty (1616-1911)

- 运茶　将采下的茶叶运来拣茶。

  Tea transportation: transport the picked tea leaves to the workshop for selection.

- 拣茶　剔除发黄、老叶茶梗等杂物。

  Tea selection: get rid of the yellow tea leaves, old tea stems, and other impurities.

- 筛茶　将烘焙好的成茶筛去碎末。

  Tea sieve: sieve away scraps from the baked tea leaves.

- 炒茶（烘茶）　利用高温迅速蒸发茶叶中的水分，保存所需色泽、味道、香气。

  Tea frying (tea baking): rapidly evaporate the moisture of the tea leaves with high temperature, preserve the desired color, taste, and aroma.

- 剔除茶梗、发黄的叶片、杂质等。

  Reject tea stems, yellow blades, and other impurities.

- 送茶　将拣好的茶叶送至晒场。
  Tea delivery: deliver the screened tea leaves to the sunning ground.

- 晒茶（萎凋）　将茶叶摊凉、蒸发，使茶叶变软，浓度加强。
  Tea sunning (withering): spread out the tea leaves for cooling and evaporation, make tea leaves soft and increase their concentration.

- 揉捻　将茶叶捣碎，使其受压而变得紧结。
  Rolling: smashing the tea leaves, compact them by pressing.

- 将发酵好的茶叶送至烘焙作坊。
  Transport the fermented tea leaves to the baking workshop.

- 发酵　可去掉茶叶的青涩味，使味道、颜色更好。
  Fermenting: get rid of the astringent taste of the tea leaves and improve their taste and color.

陆羽设计的28种茶具
The 28 Kinds of Tea Wares Designed by Lu Yu

- 灰承
  Ash Tray

- 筥
  Bamboo Basket

- 鍑
  Cauldron

- 风炉
  Stove

- 夹
  Bamboo Nipper

- 交床
  Kettle Stand

- 拂末
  Tea Broom

- 罗
  tea Sieve

- 合
  Tea Case

- 炭挝
  Coal Breaker

- 纸囊
  Paper Bag

- 碾
  Tea Roller

- 火夹
  Fire Tongs

- 碗
  Tea Bowl

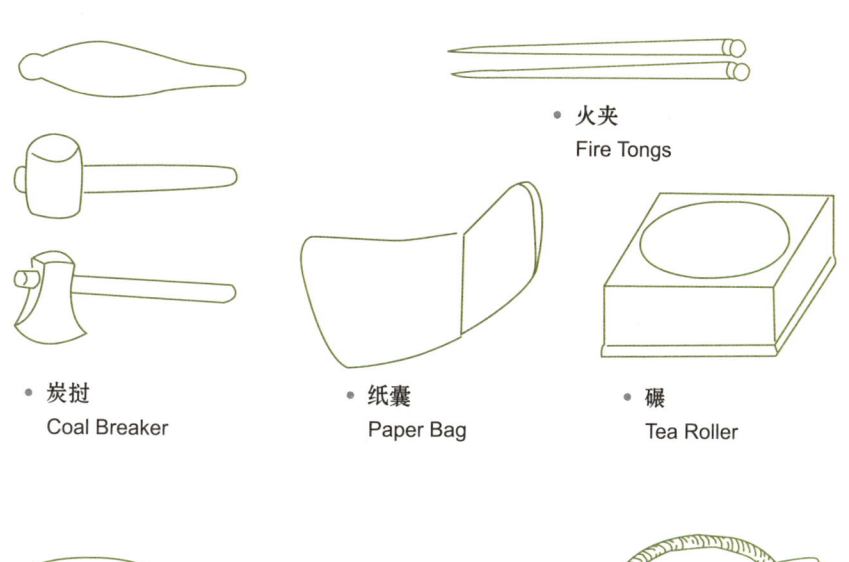

- 揭
  Opener

- 漉水囊
  Water Filter

- 具列
  Tea Set Shelves

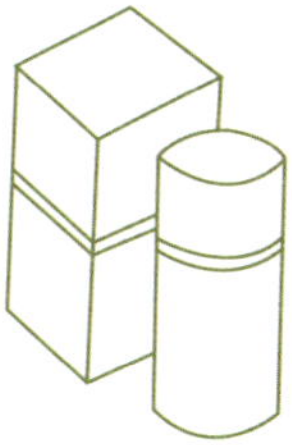

- 畚
  Tea Bowl Container

- 都篮
  Tea Wares Container

- 涤方
Rinsing Container

- 滓方
Dregs Container

- 巾
Tea Scarf

- 水方
Water Container

- 则
Tea Weighing Scoop

- 醯簋
Salt Bottle

- 熟盂
Hot Water Jar

- 瓢
Gourd Ladle

- 札
Teapot-scouring Brush

> 中国人爱饮茶

中国人的日常生活中有七样必需品：柴米油盐酱醋茶。其中只有茶是饮品，茶已经完全融入中国人的生活中，是与食物同样重要的东西。对于中国人来说，茶虽然不是饭，不能果腹，但却非喝不可。而且饮茶不仅可以解渴，还是一种生活方式，超越生理需求上升到了精神层面。中国语言学家林语堂曾说，中国人最爱品茶，在家中喝茶，开会时喝茶，打架、讲理也要喝茶，早饭喝茶，午饭后也要喝茶。中国人似乎在任何地方、任何时间都可以饮茶。饮茶是中国人奉行的生活文化。每当有客人来访，主人就会沏上一壶茶来招待客人，这种习俗一直延续到现在。在宴请

> Chinese People's Love for Drinking Tea

Chinese people have seven necessities in their daily life: firewood, rice, oil, salt, sauce, vinegar, and tea. Among them, tea is the only drink. Tea has fully merged into Chinese people's daily life and became as important as food. For the Chinese, although tea does not fill their stomach like rice, they just cannot do without it. In fact, tea can not only quench thirst, but also represents a life style. People's love of tea is more of a spiritual demand than a physical one. Lin Yutang, a famous Chinese linguist, once said that Chinese people love tea the most; they drink tea at home, during a meeting, when reasoning to solve a conflict, during breakfast, and after lunch. They seem to be able to drink tea anywhere and anytime. Drinking tea is a life culture upheld by Chinese people.

客人时，主人有时会用酒来表达敬意，而茶是唯一能代替酒的饮品，"以茶代酒"既不失礼仪，又令人愉快。茶不仅是中国人日常和宴客的饮品，也是婚礼、生日或葬礼等大型聚会时不可或缺的饮品。

When guests come, the host will always make a cup of tea to treat them. This custom has been passed down to this day. When entertaining guests with a banquet, the host will sometimes propose a toast to the guests to show his respect. Tea is the only qualified substitute for the wine on this occasion. It meets the protocol and keeps everybody happy. Tea is an indispensable drink for Chinese people in their daily life and during formal gatherings such as wedding, birthday party, or funeral.

- 《寒夜客来茶当酒》齐白石（近代）

Using Tea as a Substitute for Wine to Treat a Guest in a Winter Night, by Qi Baishi (Modern Times)

- 《宫乐图》佚名（唐）

图中描绘的是唐代宫廷仕女宴乐品茗的场景。

The Painting of a Happy Tea Banquet in the Palace, by an anonymous author (Tang Dynasty 618-907)

The painting depicts the scene of a happy tea banquet for maidservants in the palace of the Tang Dynasty (618-907).

茶俗

　　茶俗是指在长期的社会生活中，逐渐形成的一种以茶为主题或是以茶为媒介的风俗习惯。茶俗具有地域性、阶级性、演变性等特点。中国传统茶俗根据实际情况的不同可分为日常饮茶、客来敬茶、婚恋用茶、祭祀供茶等。

　　客来敬茶是一种传统的茶俗。中国是文明古国、礼仪之邦，好客的中华民族逐渐将日常茶之品饮演化为一种待客之道。以茶敬客之风最早出现于魏晋南北朝时期，唐宋时期客来敬茶的习俗普及开来，有"客来设茶，送客点汤"之说。即有客来，为客人端上一杯香茶以示尊重和欢迎；"点汤"则有"端茶送客"之意。当主人双手端起茶杯，即有逐客之意，客人不得不起身告辞。如今，"以茶待客"更是成为人们生活中最常见的一种待客形式了。

　　婚恋用茶是指在婚恋的过程中茶的使用形式，有订亲茶、受聘茶、新娘子茶，这些用茶形式在中国少数民族中最为常见。茶饼也是结婚送礼时的重要物品之一。以茶为聘礼，是一种吉祥的象征，寄托着人们对新人的美好祝愿。

Tea Custom

Tea custom refers to the social custom and habits focusing on tea or carried on by tea that have gradually taken shaped during the long-term social life. Tea custom shows different

- 《萧翼赚兰亭图》阎立本（唐）
  图中描绘的是儒生萧翼与僧人共同喝茶的场景。
  The Painting of Xiao Yi Obtaining the "Preface to the Poems Composed at the Orchid Pavilion" with a Plot, by Yan Liben (Tang Dynasty 618-907)
  The painting depicts the scene when scholar Xiao Yi had tea with the monks.

characteristics of different regions, different classes, and different periods. Traditional tea custom of China varies with actual conditions and can be divided into those kept in daily tea drinking, guest entertainment with tea, tea consumption during marriage arrangement and wedding, and sacrifice offering with tea.

Offering tea to guests is a traditional tea custom. China is a country with long-standing civilization and a land of propriety and righteousness. The hospitable Chinese people have gradually turned tea drinking from a daily practice into a refined way of receiving guests. Tea was first offered to the guests in the Wei, Jin, and Northern and Southern dynasties. Such custom became a popular practice in the Tang and Song dynasties. The old saying goes that "when a guest comes, the host offers tea to him as a gesture of respect and welcome; when the host takes up the tea cup with both hands, he urges the guest to leave." Now, "treating guests with tea" is the commonest way of entertaining guests in Chinese people's daily life.

Tea consumption during marriage arrangement and wedding includes engagement tea and bridal tea, which are the commonest ways of tea consumption among the minority ethnic groups of China. Caky tea is also one of the important gifts during wedding ceremony. Tea as a betrothal gift is an auspicious symbol that carries people's good wishes for the new couple.

中国茶馆

中国茶馆与西方咖啡馆有许多相似之处,是休闲娱乐之地。但是两者的气氛却有很大的不同,茶馆是热闹的,咖啡馆是安静的。茶馆中有很多娱乐活动,如曲艺表演、说唱评书等,人们一边欣赏着表演,一边喝着茶,度过悠闲的时光。

茶馆历史悠久,唐代文献中已有记载。宋代时茶馆已经十分普遍,出现了"斗茶"项目,这是茶馆中重要的节目。茶馆在此时也有了档次之别,高档茶馆开始讲究室内装潢。

茶馆自明代起就开始变得雅致起来,讲究水、茶、茶具等。许多文人雅士频繁进出茶馆,促进了茶馆文化品位的提升。茶馆不再仅仅是喝茶休憩的场所,其社会功能渐渐丰富起来。茶馆文化在明清时期成熟起来,到了民国时期,茶馆的阶层界限则相当分明,有的茶馆高雅,有的茶馆粗俗,适应着不同阶层茶客的需要。来茶馆的人多了，能够在茶馆里做的事也多了，特别是靠打通人际关系来谋事的商人们，都会选择在茶馆会客、洽谈买卖，茶馆成了他们的交易场所。中国人民艺术家老舍先生创作的话剧《茶馆》讲述的就是茶馆里的故事。

现代茶馆的社会功能更加丰富，可啜茗饮茶，可小憩休闲，可洽谈商务，还可体验茶文化，已经成为地方文化的一种标识。茶文化中最直接的表现形式首推茶艺，几乎所有的现代茶馆都在展示自己的茶艺。

- 茶馆里的曲艺表演
A Chinese Folk Art Performance in the Teahouse

- 茶馆里精彩的茶艺表演
The Wonderful Tea Art Performance in the Teahouse

Chinese Teahouse

Chinese teahouses have many things in common with western cafes. They are both the venue for leisure spending and amusement. However, they have totally different atmospheres. A teahouse is bustling with activities while a cafe is quiet. Among the recreational activities in a teahouse are Chinese folk art performances such as comic dialog, ballad singing, and story telling. Tea drinkers enjoy the tea and the performances in the teahouse, spending their leisure hours happily.

Teahouse has a long history in China. It was recorded in the literature of the Tang Dynasty. By the Song Dynasty, teahouse had been rather popular and an important teahouse event, Tea Contest, appeared. At that time, teahouses were at different levels. The first-class ones had exquisite interior decoration.

Since the Ming Dynasty, the teahouse became elegant with its increasing emphasis on water, tea, and tea set. Many refined scholars frequented the teahouse and boosted the upgrading of its cultural taste. A teahouse was no longer a place merely for tea and rest, but a venue with richer and richer social functions. Teahouse culture matured in the Ming and Qing dynasties. By the Minguo Period, there was a clear boundary between the teahouses serving the people from different social strata. Some teahouses were for highbrows while some others for lowbrows. With more and more people coming to the teahouse, they found it a place suitable for more and more purposes. The businessmen, in particular, chose to meet guests and hold business talks at the teahouse. The stage play, *Teahouse*, by Mr. Lao She, a well-known Chinese artist, talks about the stories happened in a Chinese teahouse several decades ago.

Modern teahouse has even richer social functions. It is a venue for people to drink tea, have a rest, hold business talks, and experience tea culture. It has become a symbol of the local culture. Tea culture is demonstrated most directly by tea art. Almost all modern teahouses are engaged in creating their own tea ceremonies.

- **话剧《茶馆》场景再现（老舍茶馆泥塑）**

  话剧《茶馆》是老舍先生1957年完成的作品。全剧以老北京一家大茶馆的兴衰变迁为背景，向人们展示从清末到抗战胜利的半个世纪里北京的社会风貌和各阶层人物的命运。该剧是中国目前演出场次最多的剧目之一，曾在法国、瑞典、日本、加拿大、新加坡等国家上演并获得好评。

  Reconstruction of a Scene of the Stage Play *Teahouse* (Clay Sculpture of a Scene of Lao She's Stage Play *Teahouse*)

  The stage play *Teahouse* was completed by Mr. Lao She in 1957. By presenting the stories of the rise and fall of a large teahouse in the old-day Beijing, it demonstrates the social conditions and the fate of the people from different social strata during the half century from the end of the Qing Dynasty to the victory of the War of Resistance Against Japanese Aggression. It is one of the plays staged most frequently in China. It has been played and well received in France, Sweden, Japan, Canada, and Singapore.

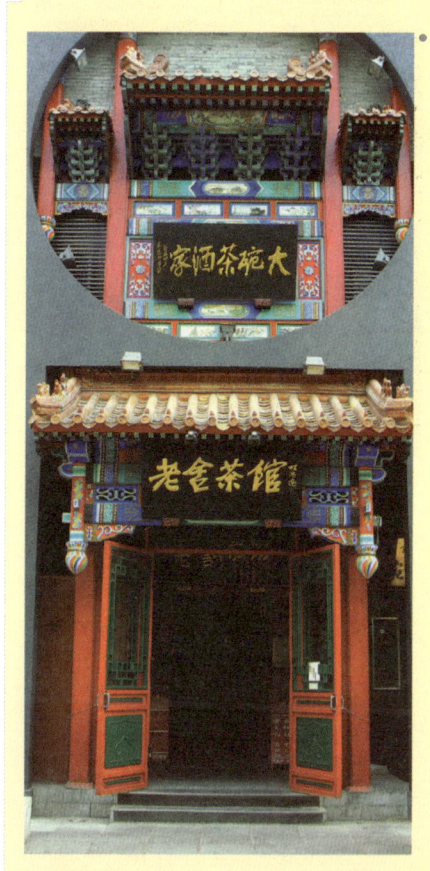

- **老舍茶馆**

老舍茶馆始建于1988年，是以老舍先生及其名著《茶馆》命名的。老舍茶馆地处前门，沿袭了明清时期茶馆的风格和特色，京味儿十足。老舍茶馆自开业以来，接待了近50位外国元首、众多社会名流和200多万中外游客，是展示北京文化的特色窗口。

老舍茶馆在形式和功能上都继承了北京老茶馆的韵味。在门口接待的是身穿蓝色大马褂儿、肩搭白色手巾板儿的小堂倌，用地道的北京腔吆喝着。大厅四壁挂有书画楹联，书香气息浓郁。内部装饰以传统中式风格为主，木质廊窗、大红宫灯、红木桌椅、细瓷盖碗，处处透露着北京的民俗风情。老舍茶馆的二楼还设有一个四合茶院，创造性地将老北京经典四合院纳入楼宇之中。老舍茶馆每晚都有各式节目演出，有京剧、变脸、中国功夫、茶艺表演、相声、双簧、快板、单弦、魔术、杂技等等。

The Lao She Teahouse

The Lao She Teahouse was built in 1988 and named after Mr. Lao She and his famous stage play *Teahouse*. Located at Qianmen, the Lao She Teahouse adopts the style and characteristics of the teahouses in the Ming and Qing dynasties. It is full of the taste of old Beijing. Since the commencement of its operation, the Lao She Teahouse has received nearly 50 foreign state leaders, many social celebrities, and over two million visitors from at home and abroad. It has become a window to display Beijing's cultural characteristics.

The Lao She Teahouse has inherited the charm and taste of the old Beijing teahouses in both form and function. Receiving the guests at the door are young waiters in blue mandarin jacket and with a white towel on the shoulder. They cry out for customers in idiomatic Beijing dialect. On the wall of the teahouse's main hall are Chinese calligraphy works, paintings, and couplets. The interior decoration of the teahouse is also in the traditional Chinese style. With wooden corridor windows, red palace lanterns, mahogany tables and chairs, and fine porcelain tureens, the teahouse displays to its visitors the genuine folk custom of Chinese tea culture. On the second floor of the teahouse is a courtyard where the tea drinkers can have a first-hand experience of the classical courtyard life style of the old Beijing. Every night, the Lao She Teahouse presents various kinds of performances, including Peking opera, face change, Chinese Kungfu, tea art performance, comic dialog, two-man comic show, clapper talk, story-telling to musical accompaniment, magic, and acrobatics.

• 顺兴老茶馆

四川省成都市的顺兴老茶馆，集明清建筑、壁雕、窗饰、木刻、家具、茶具、服饰和茶艺于一体，是传承巴蜀文化的经典杰作。

顺兴老茶馆将怀旧做到了极致。老茶馆内所用的一砖一石，都是从蜀都城乡的千家万户中搜集而来，全部都是历经沧桑的古砖古石。另外，桌椅木窗、茶壶茶碗、房檐木匾等都是旧物。整个茶馆里弥漫着浓浓的怀旧情愫。

The Shunxing (Smooth and Prosperous) Old Teahouse

Located in Chengdu City of Sichuan Province, the Shunxing Old Teahouse demonstrates many elements of the society in the Ming and Qing dynasties, including the architecture, mural carving, window decoration, wood carving, furniture, tea set, costumes, and tea art, making itself a classical masterpiece in inheriting Sichuan culture.

The Shunxing Old Teahouse has made extreme efforts in putting on a nostalgic atmosphere. The bricks and stones in the teahouse were collected from many urban and rural households in the province. They bear testimony to the vicissitudes of the time. Besides, the tables, chairs, wooden windows, teapots and tureens, eaves and wooden boards of the teahouse are all old things. The entire teahouse is covered in a thick nostalgic atmosphere.

• 长嘴壶茶艺表演

在顺兴老茶馆还能欣赏到长嘴壶茶艺。"茶博士"手持壶嘴有一米多长的大铜壶，在翻转腾挪之后，不用走到顾客身边，隔着桌子就能稳稳地把滚水注入杯盏之中，做到不洒、不漏、不溢、不滴，堪称绝活。

Tea Art Performance with Long-spout Teapot

At the Shunxing Old Teahouse, the tea drinkers can enjoy tea art performance by waiters armed with long-spout teapots. Teahouse waiters patrol the teahouse carrying large teapots with a one-meter-long spout. They make their way through the crowded teahouse smoothly and pour water into the waiting teacups gracefully. They do not have to get close to the drinkers in need. The long spout enables them to pour the hot water from afar. They never miss the target and never waste a single drop of water. The performance is an attraction in its own right.

名山出好茶
Good Teas from Great Mountains

  在中国广大的产茶区域内，但凡有名山的地方，几乎都产有名茶。名山，有葱郁的森林，如烟的云海，长流的清溪，正适合喜温、喜湿、耐阴的茶树生长。中国的历史名茶和当今新创制的名茶，几乎都产于名山之上。

China has vast tea-producing areas. Almost all the great mountains in the country produce famous teas. The great mountains have lush forests, misty cloud seas, and clear springs. Together, they make a warm, humid, and shady environment suitable for the growth of tea trees. Almost all the famous teas made in the past and modern times are from the great mountains.

> 武林山龙井茶

　　杭州，古称"武林"。杭州西湖东面濒城，三面环山，即西山、南山和北山，统称"武林山"。在武林山水间，名胜古迹颇丰，著名的有"西湖十景"：断桥残雪、平湖秋月、曲院风荷、苏堤春晓、双峰插云、花港观鱼、三潭印月、南屏晚钟、雷峰夕照、柳浪闻莺。

> Longjing (Dragon Well) Tea from Mt. Wulin

Hangzhou was known in the ancient times as Wulin. The West Lake of Hangzhou faces the city in the east and is encircled by mountains on the other three sides. The western, southern, and northern mountains are called Mt. Wulin as a whole. Between the lake and the mountains are many famous historic and cultural sites. The most famous are the ten views of the West Lake, namely the Melting Snow at the Broken Bridge, Moon over the Calm Lake in Autumn, Wine-making Yard and Lotus Pool in Summer, Dawn on the Su Causeway in spring, Twin Peaks Piercing the Clouds, Viewing Fish at the Flower Pond, Three

• 西湖风光
Landscape of the West Lake

龙井，是井名、寺名，也是村名、茶名，而以龙井茶最为人称颂。西湖龙井茶生长环境得天独厚，又加上早期广福院、灵隐寺、天竺寺、净慈寺、韬光寺和道朴院等寺庙得道僧侣们的悉心培制，终于成为"天下第一名茶"。北宋诗人苏东坡曾写诗赞誉西湖龙井茶之美，称其犹如美女西施一般空后绝前。

Ponds Mirroring the Moon, Evening Bell at the Nanping Hill, Sunset Glow at Leifeng Pagoda, and Orioles Singing in the Willows.

Longjing is the name of a well, a temple, a village, and also a tea. It is best known as the name of a tea. The West Lake Longjing Tea is hailed the No.1 Tea in the world thanks to the unique growing environment around the West Lake and the early painstaking cultivation

- 龙井之乡
  Hometown of the Longjing Tea

• 西湖龙井茶园

西湖一带产茶历史悠久,早在陆羽《茶经》中就记述了杭州天竺和灵隐二寺产茶。

A West Lake Longjing Tea Garden

The West Lake area has long been known as a tea-producing place. Lu Yu recorded in his *The Classic of Tea* the teas produced by the Tianzhu Temple and Lingyin Temple in Hangzhou City.

龙井茶的核心产区集中于狮峰山、梅家坞、翁家山、云栖、虎跑、灵隐等地。宋代僧人辩才禅师在狮峰山麓首开种茶先例,又常与苏东坡、秦观等文人品茶吟诗,故龙井茶之名始见于诗文之中。时至今日,狮峰山龙井茶仍被认为是龙井茶中品质最高的,辩才禅师则被尊称为"龙井茶祖"。

产于狮峰山一带的龙井称为"狮"字号龙井。此外,还有"龙"字号、"云"字号、"虎"

by the monks of several temples in the area, including the Guangfu Temple, Lingyin Temple, Tianzhu Temple, Jingci Temple, Taoguang Temple, and Daopu Temple. Su Dongpo, the famous poet in the Northern Song Dynasty (960-1127), once wrote a poem to praise the beauty of the West Lake Longjing Tea. He likened the tea to the top Chinese beauty Xi Shi and regarded both as unprecedented and unrepeatable.

The Longjing Tea is produced mainly in Mt. Shifeng, Meijiawu, Mt.

字号、"梅"字号龙井。"龙"字号产于翁家山、杨梅岭、满觉陇、白鹤峰一带,本地人称"石屋四山"龙井;"云"字号产于云栖、五云山、琅铛岭西等地;"虎"字号传统产地为虎跑、四眼井、赤山埠、三台山一带,后来,法云弄、白乐桥、玉泉、金沙港、茅家埠、黄龙洞一带茶园的茶叶也被划为"虎"字号龙井;"梅"字号为梅

- **狮峰山**
  以狮峰山为中心产地的"狮"字号龙井为西湖龙井茶中的"绝品"。
  Mt. Shifeng (Lion Peak)
  S-Longjing produced in the area centering on Mt. Shifeng is the top product of the West Lake Longjing Tea.

Wengjia, Yunqi, Hupao, and Lingyin. In the Song Dynasty (960-1279), Zen master Biancai took the lead to plant tea in Mt. Shifeng. He often held tea party with famous poets Su Dongpo and Qin Guan. They enjoyed tea and chanted poems together, hence the entry of the name of the Longjing Tea into the poems. Today, the Longjing Tea from Mt. Shifeng is still regarded as the best. Zen master Biancai is also respectfully called the Forefather of the Longjing Tea.

The Longjing Tea produced in Mt. Shifeng is known as S-Longjing. There are also L-Longjing, Y-Longjing, H-Longjing, and M-Longjing. L-Longjing is produced in Mt. Wengjia, Mt. Yangmei, Mt. Manjue, and Mt. Baihe, which are called the Four Rock-shed Mountains by the locals. Y-Longjing is produced in Yunqi, Mt. Wuyun, and west of Mt. Langdang. H-Longjing was traditionally produced in Hupao, Siyanjing, Chishanbu, and Mt. Santai. Later, the teas produced in Fayunnong, Baileqiao, Yuquan, Jinshagang, Maojiabu, and Huanglongdong are also classified as H-Longjing. M-Longjing is produced in the area around Meijiawu.

Picking of the Longjing Tea focuses on three elements, namely Early Picking,

• 十八棵御茶

相传这十八棵茶树是由清乾隆皇帝所封。他曾四次到狮峰山下胡公庙品饮龙井茶，饮后赞不绝口，兴之所至，便将庙前十八棵茶树封为"御茶"。

Eighteen Imperial Tea Trees
According to the legend, these eighteen tea trees were entitled the Imperial Tea Trees by Emperor Qianlong in the Qing Dynasty. The emperor visited Mt. Shifeng four times and he was pleased by the Longjing tea he was offered at the Hugong Temple on the foot of Mt. Shifeng. In high spirits, he conferred imperial title on the eighteen tea trees.

家坞一带所产。

龙井茶的采摘要求是"早、嫩、勤"。"早"是指采摘的时间，清明前采制的龙井茶品质最佳，称"明前茶"，谷雨前采摘的品质尚好，称"雨前茶"。"嫩"是指采摘的芽叶要细嫩完整，标准为一芽一叶及一芽二叶初展。"勤"是指采摘的量，每公斤干茶需7~8万个鲜嫩芽叶。

Picking of Tender Leaves, and Thorough Picking. The tea picked before the Qingming Festival, or Pure Brightness, one of the 24 solar terms in Chinese lunar calendar, has the best quality and is called Pre-PB tea. The tea picked before Grain Rain, another solar term, has a good quality and is called Pre-GR tea. Picking of Tender Leaves refers to the picking of the tender and complete tea buds and leaves, usually the first bud with the

龙井嫩芽
Tender Buds of the Longjing Tea

龙井茶素以"色绿、香郁、味甘、形美"著称于世。通常通过外形、香气、滋味、汤色和叶底五个方面来鉴赏龙井茶的品质。外形以扁平光滑、挺秀尖削、均匀整齐、色泽翠绿鲜活为上品；品质好的龙井茶带有鲜纯的嫩香，香气清醇持久；滋味与香气成正比，香气好的茶叶，通常滋味也好，以鲜醇甘

first one or two leaves starting to open. Thorough Picking refers to the quantity of the picked tea leaves. Each kilogram of dry tea contains 70,000 to 80,000 fresh tender tea buds and leaves.

The Longjing Tea has been known for its green leaves, rich aroma, sweet taste, and beautiful shape. Normally, the quality of the Longjing Tea is judged on five aspects, namely appearance, aroma, taste, soup color, and brewed leaves. The best-quality tea has a flat and smooth appearance, tapering leaves, an even and neat contour, and a fresh green color. The good-quality Longjing Tea has a pure and fresh aroma that lasts long. Its taste is in direct proportion to its aroma. The tea leaves with good aroma usually have good taste. The fresh sweet taste is the best. Tea soup is judged on its tone, brightness, and clearness. The

---

清明、谷雨

清明：中国农历二十四节气之一，在每年的4月4日至6日之间。
谷雨：中国农历二十四节气之一，在每年4月19日至21日之间。

Pure Brightness, Grain Rain

Pure Brightness: one of the 24 solar terms in Chinese lunar calendar, between April 4th and 6th
Grain Rain: one of the 24 solar terms in Chinese lunar calendar, between April 19th and 21st.

爽为好；汤色主要看色度、亮度和清浊度，以清澈明亮为好；优质的龙井茶叶底芽叶细嫩成朵、大小匀齐、嫩绿明亮。

clear and bright tea soup is the best. The high-quality Longjing Tea has tender, complete, green, and bright brewed leaves that are in even sizes.

• **西湖龙井茶样、西湖龙井茶汤**
西湖龙井属嫩扁形炒青绿茶。
Sample of the West Lake Longjing Tea; Soup of the West Lake Longjing Tea
The West Lake Longjing Tea is a stir fixation green tea with a tender flat appearance.

绿茶

绿茶，是人类制茶史上出现最早的加工茶，属于不发酵茶。其最显著的特征是清汤绿叶，即干茶色泽绿，茶汤黄绿明亮，叶底（茶渣）鲜绿。

绿茶的品种最多、产量最高、产区最广，几乎中国的各产茶区均产绿茶。

绿茶制作一般经过杀青、揉捻、干燥三道工序。其中杀青是决定绿茶色泽的关键工序，即用高温破坏茶叶中酶的活性，制止茶多酚类物质的酶促氧化作用，使叶中水分蒸发，青臭气发散出去，以产生茶香。

根据杀青方式和最终干燥方式不同，绿茶分为蒸青绿茶、炒青绿茶、烘青绿茶和晒青绿茶四类，还有介于蒸青和炒青之间的半烘炒绿茶。

绿茶由于未经发酵，较多地保留了鲜茶叶中的天然成分，其中茶多酚和咖啡碱保留了85%以上。绿茶中的茶多酚是由多种酚类物质组成的复合物，主要包括儿茶素、黄酮素、花青素和若干酚酸。其中儿茶素含量最高，约占茶多酚总量的70%以上，能够增强毛细血管的活性、预防动脉硬化、止泻、杀菌、消炎等。另外绿茶中还含有丰富的氨基酸和人体需要的大量元素和微量元素，包括磷、钙、钾、钠、镁、硫、铁、锰、锌、硒、铜、氟、碘等。这些元素对保持人体生理健康能起到重要的调节作用。

- 绿茶的品质特点就是"三绿"，即叶绿、汤绿、叶底绿。
The Quality Features of the Green Tea are Three Greens, Namely Green Leaves, Green Soup, and Green Brewed Leaves.

绿茶虽然营养成分高，但并不是所有人都适合饮用。中医认为绿茶性偏寒，所以不适合发烧患者、肝脏病人、神经衰弱者、怀孕期妇女、哺乳期妇女、溃疡病患者和营养不良者饮用。

绿茶中的名品很多，有西湖龙井茶、洞庭山碧螺春、庐山云雾茶、黄山毛峰、太平猴魁、峨眉竹叶青、蒙顶甘露茶、信阳毛尖茶、顾渚紫笋茶、普陀佛茶、安吉白茶、都匀毛尖茶、南京雨花茶等等。

Green Tea

Green tea is the earliest unfermented tea processed by human. Its most obvious features are the green soup and green leaves. The dry tea leaves are green and the tea soup is bright yellowish green. The brewed tea leaves (tea dregs) are fresh green.

The green tea has the most varieties. Its output is the highest among all teas. It is produced in the widest areas. Almost all tea-producing areas in China produce green tea.

Producing green tea will normally go through three processing procedures: fixation, rolling, and drying. Among them, fixation is the key procedure to decide the color of the green tea. During fixation, high temperature is applied to destroy the activity of the enzyme in the tea and stop the enzymatic oxidation of green tea polyphenol (GTP) substances. Through fixation, the tea leaves are dehydrated and their green odor is dispersed. As a result, tea aroma is created.

Based on fixation modes and final drying modes, green tea can be divided into four types:

steamed green tea, stir fixation green tea, baked green tea, and sunned green tea. There is also the semi-baked and semi-stir-fixation green tea.

As green tea is not fermented, it contains large amount of the natural components of the fresh tea leaves, including over 85% of the GTP and caffeine. GTP in green tea is a compound composed of several phenol substances, mainly including catechin, flavone, anthocyanidin, and several phenolic acids. Among them, catechin content is the highest, accounting for over 70% of GTP. Catechin can enhance the activity of capillary vessel, prevent arteriosclerosis, stop diarrhea, kill bacteria, and reduce inflammation. In addition, green tea also contains rich amino acid and many elements and trace elements needed by human body, including phosphorus, calcium, potassium, sodium, magnesium, sulfur, iron, manganese, zinc, selenium, copper, fluorine, and iodine. These elements play an important regulating role in keeping physiological health of human body.

Though containing high content of nutrients, green tea is not necessarily suitable for all people. Traditional Chinese medicine regards green tea as cold in nature and is not suitable for the patients suffering from fever, liver diseases, neurosism, ulcer, and malnutrition, and women during gestation and lactation periods.

There are many famous green tea varieties, including the West Lake Longjing Tea, Biluochun (Green Spiral Spring) Tea from Mt. Dongting, Yunwu (Clond and Mist) Tea from Mt. Lushan, Maofeng (Hairy Tip) Tea from Mt. Huangshan, Houkui Tea from Taiping, Zhuyeqing (Green Bamboo Leaf) Tea from Mt. Emei, Ganlu (Sweet Dew) Tea from Mengding, Maojian (Hairy Tip) Tea from Xinyang, Zisun (Purple Bamboo) Tea from Mt. Guzhu, Buddha Tea from Mt. Putuo, White Tea from Anji, Maojian (Hairy Tip) Tea from Duyun, and Yuhua (Rain) Tea from Nanjing.

• 绿茶中的多种成分有益人体健康
Many Contents in Green Tea are Conducive to Human Health

> 顾渚山紫笋茶

顾渚山位于浙江省湖州市长兴县西北部，属水口乡顾渚村，产茶历史悠久，出产的紫笋茶最为著名。顾渚山海拔355米，西北紧临黄龙头、啄木岭、悬臼山、乌头山等天目山余脉低山丘陵，形成阻挡冬季寒流的天然屏障；东面距太湖10千米，春夏时节多东南风，太湖水面暖湿的空气被带进山谷。野生茶树就生长在这些微域气候条件较好的山谷中，主要分布在顾渚村的方（桑）坞、高坞、竹坞、狮坞、斫射山（古称明月峡）。这里植被丰茂，为喜阴怕晒的茶树提供了荫蔽。此外，山中土壤为乌沙土，氮、磷、钾等有机质含量高，土层深厚。上述这些天然因素，为紫笋茶品质的形成奠定了基础。

> Zisun (Purple Bamboo Shoot) Tea from Mt. Guzhu

Mt. Guzhu stands in the northwest of Changxing County, Huzhou City, Zhejiang Province. The area belongs to Guzhu Village of Shuikou Township. Tea has long been produced here. Of local teas, the Zisun Tea is the most famous. Mt. Guzhu is 355 meters above sea level. It adjoins on some low hills of Mt. Tianmu in northwest, including Mt. Huanglongtou, Mt. Zhuomuling, Mt. Xuanjiu, and Mt. Wutou, forming a natural barrier to stop winter cold waves. It is ten kilometers from the Taihu Lake to the east. In spring and summer, the area is subject to southeast wind, which brings humid air from the Taihu Lake into the mountain valleys. Wild tea trees are growing in these valleys with good micro climate conditions, mainly including Fangwu (Fangsangwu),

Gaowu, Zhuwu, Shiwu, and Mt. Zhuoshe (known as Mingyue Valley in the ancient times) of Guzhu Village. The area has lush vegetation and offers favorable shady cover for the tea trees. In addition, the mountains are covered with a thick layer of dark sand soil that contains rich organic substances such as nitrogen, phosphorus, and potassium. These natural conditions lay a solid foundation for the formation of the good quality of the Zisun Tea.

The Zisun Tea from Mt. Guzhu derives its name from its purple buds and leaves and bamboo-shoot-like backward rolling tender leaves. As far back as the Tang Dynasty (618-907), the Zisun Tea had been the favorite tribute tea of the royal family. It remained a tribute tea for 876 years through the following dynasties. In addition, the administrations of Huzhou and Changzhou prefectures in the Tang Dynasty (618-907) set up a Jinghui Pavilion in Mt. Guzhu, which served as a venue for officials from both prefectures to taste the new tea of each year. The tea-tasting party has become an event frequently talked about among tea drinkers. Not far from the Tribute Tea Station in Mt. Guzhu is the Jinsha Spring. Water from the spring was also

• 顾渚紫笋茶园
Zisun (Purple Bamboo Shoot) Tea Garden in Mt. Guzhu

　　顾渚山紫笋茶因芽叶微紫、嫩叶背卷似笋壳而得名。早在唐代，紫笋茶就成为皇家贡茶的"最爱"，后历朝续贡达876年。此外，唐时湖州和常州官府专在顾渚山上设置境会亭，每到茶季，两州官员聚集境会亭品尝新茶，成为茶坛佳话。离顾渚山贡茶院不远处的金沙泉，是唐时的贡水，与贡茶紫笋并列"双绝"。唐时顾渚紫笋是饼茶，宋时为龙团茶，明后改以芽茶进贡。现今的紫笋茶形态大为改变，根据芽叶大小分开采摘，形成了紫笋、旗芽、雀舌等品类，以紫笋为上。

顾渚紫笋是半烘炒型绿茶，于清明前至谷雨期间采制，标准为一芽一叶或一芽二叶初展。其成茶芽挺叶长，形似兰花，色泽翠绿，银毫明显，冲泡后汤色清澈明亮，滋味甘醇鲜爽，叶底细嫩成朵。

• 顾渚紫笋茶
Zisun (Purple Bamboo Shoot) Tea from Mt. Guzhu

presented to the royal family in the Tang Dynasty (618-907) together with the Zisun Tea as two tributes from the area. The Zisun Tea was a caky tea in the Tang Dynasty (618-907), a dragon lump tea in the Song Dynasty (960-1279), and a bud tea since the Ming Dynasty (1368-1644). Current Zisun Tea has a totally different appearance. According to the size of the buds, there are several varieties such as Zisun Tea, Qiya (Flag Bud) Tea, and Queshe (Sparrow Tongue) Tea. Among them, Zisun Tea is the best.

The Zisun Tea from Mt. Guzhu is a semi-baked and semi-stir-fixation green tea. It is picked and made during the period between Pure Brightness and Grain Rain when the first bud and the first one or two leaves begin to open. The mature tea has long buds and leaves, which are shaped like orchid in bright green and with obvious silver tips. The tea soup is bright and clear, with a sweet and fresh taste. The brewed tea leaves are tender and cluster into lumps.

> 普陀山佛茶

普陀山地处浙江省舟山市舟山群岛东南部，为我国佛教四大圣地之一。普陀山人文景观颇多，寺院、庵堂、茅棚就达数百处，其中，以普济禅寺、法雨禅寺、慧济禅寺较为著名。

> Buddha Tea from Mt. Putuo

Mt. Putuo is located to the southeast of the Zhoushan Archipelago of Zhoushan City, Zhejiang Province. It is one of the four Buddhist holy lands in China. Mt. Putuo has many manmade landscapes, including several hundred temples, monasteries, and thatched huts. Among them, the Puji Temple, Fayu Temple, and Huiji Temple are more famous.

- 普陀山普济寺
Puji Temple in Mt. Putuo

普陀山僧侣历来崇尚茶禅一味。从唐代开始，普济寺和其他寺院的僧侣一道，种茶制茶，世称普陀佛茶。清代时佛茶被列为贡品，后随佛事盛衰不定，茶叶生产也时续时断，1979年后得以再次恢复。现今普陀山出产的佛茶又名"佛顶云雾"。

普陀佛茶产区遍及普陀茶山及周围朱家尖、桃花岛等地。这里属于温带海洋性气候，雨量充沛，空气湿润，林木茂盛，土壤肥沃。普陀茶山位于普陀山最高峰佛顶山山

Monks of Mt. Putuo have always been believing that tea and Zen have the same principle. Since the Tang Dynasty (618-907), the monks of the Puji Temple and other temples in the mountain have been planting and making tea, which is known as the Buddha Tea from Mt. Putuo. In the Qing Dynasty (1616-1911), the Buddha Tea was chosen as a tribute tea. Later, with the rise and fall of Buddhism in the area, tea production was also interrupted from time to time. It resumed since 1979. Now, the Buddha Tea from Mt. Putuo is also called the

● 普陀山风光 (图片提供：全景正片)
Landscape of Mt. Putuo

后，自北而西，蜿蜒绵亘，中多溪涧。茶树分布在山峰阳面及山岙避风之处，孕育出品质独特的佛茶。

普陀佛茶是半烘炒绿茶，每年清明后开始采摘，仅采一季春茶。鲜叶标准为一芽一叶或一芽二叶，佛茶成茶外形紧细卷曲，色泽绿润显毫；冲泡后汤色黄绿明亮，香气清香高雅，滋味清醇爽口，叶底软亮成朵。

● 普陀山佛茶
Buddha Tea from Mt. Putuo

Yunwu (Cloud and Mist) Tea from Mt. Foding.

The Buddha Tea producing area in Mt. Putuo includes the mountain itself and the neighboring places such as Zhujiajian and Taohua Island. The area is subject to the temperate marine climate and enjoys plenty of rainfall and humid air. It has lush woods and fertile soil. The Putuo tea mountain is behind Mt. Foding (Buddha Summit), the summit of Mt. Putuo. It extends for miles from north to west and houses many springs. The tea trees are distributed on the southern slopes and the lee side of the flat grounds in the mountain. Such natural conditions give rise to the unique-quality Buddha Tea.

The Buddha Tea from Mt. Putuo is a semi-baked and semi-stir-fixation green tea. It is picked once a year in spring after the Pure Brightness. Its fresh leaves are the first bud and the first one or two leaves. The Buddha Tea has tightly curled leaves with a green, moist, and tippy appearance. The tea soup is bright yellowish green with pleasant sweet aroma. It tastes fresh and pure. The brewed tea leaves are soft and bright and cluster into lumps.

> 洞庭山碧螺春

洞庭山，位于江苏苏州的太湖之滨。太湖是我国第三大淡水湖，湖中大小岛屿48个，连同沿湖山峰、半岛，号称"七十二峰"，并以东山、西山为最，合称为洞庭山。洞庭东山宛如巨舟伸进太湖，形成一个半岛，这里雾气迷蒙，山峦隐现，湖水连天，为风物清嘉之地；洞庭西山是一个屹立在太湖中的小岛，这里茶园遍布，各种果树遍插其间，以生产中国十大历史传统名茶——洞庭碧螺春而名扬四海。

洞庭山产茶，唐时已经出名，宋时碧螺春已经成为贡茶。碧螺春茶有异香，当地人称之为"吓煞人香"。清代康熙皇帝游太湖时，有人敬献此茶。康熙皇帝十分喜欢，

> Biluochun (Green Spiral Spring) Tea from Mt. Dongting

Mt. Dongting stands by the Taihu Lake beside Suzhou City of Jiangsu Province. The Taihu Lake is the third largest freshwater lake of China. There are 48 isles of different sizes in the lake. They and the peaks and peninsulas around the lake are called Seventy-two Peaks. Among them, the Eastern Hill and Western Hill are the most conspicuous and are referred to together as Mt. Dongting. The Eastern Mt. Dongting looks like a huge boat protruding into the Taihu Lake and forms a peninsula. Covered with dense fog, its mountain range appears now and disappears then. The lake is so huge that its water touches the sky. It is indeed a beautiful place that promises good products. The Western Mt. Dongting is actually a small isle in the Taihu Lake.

● 太湖夕照
The Taihu Lake under the Setting Sun

● 洞庭碧螺春茶园
Biluochun Tea Garden in Mt. Dongting

但觉得"吓煞人香"的茶名不雅，便根据茶的色泽澄绿如碧，外形蜷曲如螺，采于早春，赐名为"碧螺春"。

It is full of tea gardens and various fruit trees. It is known at home and abroad for its production of the Biluochun Tea, one of the top ten famous historical teas of China.

Mt. Dongting has been a famous tea-producing place since the Tang Dynasty (618-907). In the Song Dynasty (960-1279), the Biluochun tea had been made a tribute tea. Due to its outstanding aroma, the Biluochun Tea was called the Scaring Aroma by the locals. Emperor Kangxi of the Qing Dynasty once paid a visit to the Taihu Lake and was offered the Biluochun Tea. He liked it very much, but regarded the name, Scaring Aroma, as inappropriate. As the tea had a green color (*Bi* in Chinese) and a spiral shape (*Luo* in Chinese) and was picked in spring (*Chun* in Chinese), the Emperor named it Biluochun.

The Biluochun Tea should be picked early as tender buds and leaves and sorted out cleanly. It is picked from March

- 《洞庭东山图》赵孟頫（元）
The Painting of the Eastern Mt. Dongting, by Zhao Mengfu (Yuan Dynasty 1206-1368)

碧螺春的采摘有三大特点：一是摘得早，二是采得嫩，三是拣得净。每年3月18日前后开采，谷雨前结束，以春分至清明采制的明前茶品质最为上等。通常采一芽一叶初展，形似雀舌，采回的芽叶不能放置，须及时进行精心挑拣，以保持芽叶匀整一致。

碧螺春茶以形美、色艳、香浓、味醇"四绝"闻名中外。其品质特点是：条索纤细，卷曲成螺，披毫隐翠，香气浓郁，滋味鲜醇甘厚，汤色碧绿清澈，叶底嫩绿明亮。另外，由于碧螺春茶树与桃树、李树、杏树、梅树、石榴等果木交错种植，茶树、果树枝桠相连，根脉相通，使得茶吸果香，花窨茶味，遂有了碧螺春茶所特有的花香果味的品质特点。

18 to the end of Grain Rain each year. The Pre-PB pickings from the Spring Equinox to Pure Brightness have the best quality. Normally, the first bud and first leaf are picked when they begin to open in the shape of a sparrow's tongue. The picked buds and leaves should not be left unprocessed. They should be timely and meticulously sorted out to keep the buds and leaves in the same size and shape.

The Biluochun Tea is well-known at home and abroad for its four extreme beauties, namely appearance, color, aroma, and taste. The tea strips are thin and curl up as spirals. Its green color is obscured by its apparent tips. The tea has strong aroma and fresh and sweet taste. The tea soup is green and clear and the brewed tea leaves are tenderly green and bright. In addition, as the Biluochun tea trees are planted amidst several fruit trees such as peach tree, plum tree, apricot tree, prune tree, and pomegranate tree, their branches get intertwined and the Biluochun Tea absorbs the fruit scent and flower fragrance.

- 碧螺春茶样、茶汤
  碧螺春属螺形炒青绿茶。
  Sample and Soup of the Biluochun Tea
  The Biluochun Tea is a stir fixation green tea with a spiral shape.

> 黄山毛峰茶

　　黄山位于安徽黄山市境内，为世界著名风景胜地，更是我国著名茶区。传说，此山乃是黄帝修身炼丹之处，故名"黄山"。黄山产茶历史悠久，名茶尤其众多，以黄山毛峰最为人们熟知。

- 黄山毛峰茶园
  Maofeng Tea Garden in Mt. Huangshan

> Maofeng (Hairy Tip) Tea from Mt. Huangshan

Mt. Huangshan is located in Huangshan City of Anhui Province. It is a world famous scenic resort and a famous tea-producing area in China. According to the legend, Yellow Emperor once practiced self cultivation and alchemy in the mountain, hence the name of the mountain as Huangshan (yellow mountain in Chinese). Mt. Huangshan has produced tea since the ancient times. Among the many famous teas it produces, the Maofeng Tea is the most well-known.

　　Mt. Huangshan has the highest peaks in the east of China. It extends 250 kilometers from east to west. Within Mt. Huangshan scenic resort, the areas around the Taohua (Peach Blossom) Peak, Ziyun (Purple Could) Peak, Yungu (Cloudy Valley) Temple, Songgu (Pine Valley) Temple, Diaoqiao (Suspension Bridge)

黄山是中国东部的最高山峰，山脉东西走向，绵延250千米。黄山风景区境内海拔700~800米的桃花峰、紫云峰、云谷寺、松谷庵、吊桥庵、慈光阁一带为特级黄山毛峰产地。风景区外围的汤口、岗村、杨村、芳村也是黄山毛峰的重要产地，历史上曾称之为"四大名家"。这里山高谷深，峰峦叠翠，溪涧遍布，森林茂密，气候温和，雨量充沛。优越的生态环境，为黄山毛峰自然品质风格的形成创造了良好的条件。

黄山产茶，历史久远。黄山毛峰前身叫"云雾茶"，黄山毛峰之名源于清光绪初年（1875），当时

• 黄山毛峰茶样、茶汤

Sample and Soup of the Maofeng Tea from Mt. Huangshan

Temple, and Ciguang (Mercy Light) Pavilion produce superfine Maofeng Tea from Mt. Huangshan. These areas are 700 to 800 meters above sea level. Some other areas outside the scenic resort are also the important producing areas for Maofeng Tea from Mt. Huangshan. These include Tangkou, Gangcun, Yangcun, and Fangcun. They were once called Top Four Tea Producers in history. These areas have high mountains, deep valleys, endless mountain peaks, crisscrossing springs, and lush forests. They enjoy a mild climate, plenty of rainfall, and an excellent ecological environment. These are the good conditions responsible for the superb natural quality of the Maofeng Tea from Mt. Huangshan.

Mt. Huangshan has long been a tea-producing place. The Maofeng Tea from Mt. Huangshan used to be called the Yunwu (Cloud and Mist) Tea. Its present name originated in the first year of the reign of Guangxu in the Qing Dynasty (1875). At that time, Xie Zheng'an, a tea trader from Shexian County, established a Xieyu (Xie's Wealth) Tea Firm. To meet market demand, he personally led his men to some famous tea gardens at Chongchuan and Tangkou in Mt. Huangshan to

歙县茶商谢正安开办谢裕大茶行，为迎合市场需求，清明前后亲自率人到黄山充川、汤口等高山名园选采肥嫩芽叶，经过精细炒焙，创制出风味俱佳的优质茶。由于该茶白毫披身，芽尖似峰，故取名"毛峰"，后再冠以地名为"黄山毛峰"。现今黄山毛峰的生产已扩展到黄山山脉南北麓的黄山市徽州区、黄山区及歙县、黟县等地。

黄山毛峰是条形烘青绿茶。从清明到立夏均为采摘期，特级黄山毛峰的采摘标准为一芽一叶初展。

pick the fat and tender tea buds and leaves around the Qingming Festival (Pure Brightness) each year. Through painstaking preparation, he produced a high-quality tea with excellent smell and taste. This tea has white tips and peak-like buds, hence its name Maofeng (Hairy Tip). Adding the name of its producing place, it is called Maofeng Tea from Mt. Huangshan. Now, Maofeng Tea from Mt. Huangshan is produced in a wider area than before, including Huizhou District, Huangshan District, Shexian County, and Yixian County of Huangshan City on the southern and northern feet of Mt. Huangshan.

The Maofeng Tea from Mt. Huangshan is a baked green tea in strip

● 黄山风光
Landscape of Mt. Huangshan

采来的芽头和鲜叶要进行选剔,使芽叶匀齐一致。黄山毛峰成茶外形似雀舌,白毫显露,色如象牙,鱼叶金黄;冲泡后汤色清澈,滋味鲜浓,醇厚甘甜,叶底嫩黄,肥壮成朵。泡好的黄山毛峰芽叶直竖悬浮,继而徐徐下沉,即使茶凉,仍有余香,人称"冷香"。

shape. It is picked during the period from Pure Brightness to the Beginning of Summer. The superfine Maofeng Tea from Mt. Huangshan is picked when the first bud and leaf begin to open. The picked buds and fresh leaves should be carefully sorted out to make them even and consistent with each other. The finished Maofeng Tea from Mt. Huangshan has the shape of a sparrow's tongue, exposed white tips, the color of the ivory, and golden leaves. The tea soup is clear and tastes fresh and sweet. The brewed tea leaves are yellow and fat and cluster into lumps. After thorough brewing, the tea leaves first stand vertically in the water and then slowly sink. Even when the tea water gets cold, it keeps the aroma, known as Cold Aroma.

- 《庐山图》张大千【局部】(近代)
The Scroll Painting of Mt. Lushan, by Zhang Daqian[Part] (Modern Times)

> 庐山云雾茶

庐山，位于江西九江市南，屹立于长江之畔，独秀在鄱阳湖之滨，为全国重点风景名胜区，并被列入《世界自然遗产名录》。庐山是一座地垒式断块山，素以"雄、奇、险、秀"闻名于世。群峰间散布冈岭26座，壑谷20条，岩洞16个，怪石22处，瀑布22处，溪涧18条，湖潭14处。著名的三叠泉瀑

> Yunwu (Cloud and Mist) Tea from Mt. Lushan

Mt. Lushan is located in the south of Jiujiang City of Jiangxi Province. It stands by the Yangtze River and the Poyang Lake. It is a key national scenic resort and has been included in the *List of World Natural Heritages*. Mt. Lushan is a horst block mountain and has been known for its majesty, uniqueness, precipitous terrain, and elegance. It has 26 peaks,

布，落差达155米。庐山上五峰耸立，犹如五个老人并坐，人称"五老峰"。从各个角度仰望远视，由于山姿不一，有像诗人吟咏的，有像勇士高歌的，有像渔翁垂钓的，有像老僧盘坐的。五老峰青莲寺，还曾是唐代诗人李白的隐居之地。

东汉时，庐山成为中国佛教中心之一，出现了三大名寺——东林寺、西林寺、大林寺。庐山云雾茶的前身庐山茶，即为东晋时东林、西林、大林等古刹名寺的僧侣种植和创制。到唐代时，庐山茶已很著名，云雾茶之名则是出现在明代《庐山志》之中。

20 valleys, 16 grottos, 22 grotesque rocks, 22 waterfalls, 18 springs, and 14 lakes and pools. The famous Three-step Waterfall has a drop of 155 meters. The five major peaks [Wulao (Five Eldly Men) Peak as a whole] of Mt. Lushan tower into sky like five old men sitting side by side. Viewed from different angles, they are like different figures. This one is like a chanting poet and that one a singing warrior. This one is like an angling fisherman and that one a sitting old monk. The Qinglian (Green Lotus) Temple in the mountain was once the hermitage of Li Bai, a famous poet in the Tang Dynasty (618-907).

● 庐山云雾茶
Yunwu Tea from Mt. Lushan

庐山云雾茶的采摘以一芽一叶初展为标准，经杀青、抖散、揉捻、炒二青、理条、搓条、拣剔、提毫、烘干九道工序制成。成茶条索紧结重实，饱满秀丽，色泽碧嫩光滑，茶芽隐绿；冲泡后汤色明绿，香气芬芳、高长、锐鲜，滋味鲜爽醇甘，叶底嫩绿微黄，柔软舒展。

- 冲泡后的庐山云雾茶
  The Brewed Yunwu Tea from Mt. Lushan

In the Eastern Han Dynasty (25-220), Mt. Lushan became one of the Buddhist centers of China. Three famous temples appeared in the mountain and they were Donglin (Eastern Woods) Temple, Xilin (Western Woods) Temple, and Dalin (Big Woods) Temple. The Yunwu Tea from Mt. Lushan was known as Lushan Tea before. It was planted and made by monks of the three temples in the Eastern Jin Dynasty (317-420). By the Tang Dynasty (618-907), the Lushan Tea had been very famous. The Yunwu Tea had appeared in the *Chronicles of Mt. Lushan* in the Ming Dynasty (1368-1644).

The Yunwu Tea from Mt. Lushan is picked when the first bud and first leaf begin to open. Its production requires nine procedures: fixation, loosing up, rolling, stir fixation, sorting out, rubbing, sorting out, tipping, and baking. The finished teas are tight, full, and elegant strips with a green and smooth appearance. The tea bud is vaguely green. The tea soup is bright and green with strong, acute, and lasting aroma. It tastes sweet and brisk. The brewed tea leaves are greenish yellow, soft, and opening.

• 五老峰下的云雾茶茶园
A Yunwu Tea Garden under the Wulao Peak

• 庐山风光
Landscape of Mt. Lushan

> 峨眉山竹叶青茶

峨眉山，位于四川峨眉市西南，为全国重点风景名胜区，是中国佛教四大圣地之一。山中名胜古迹颇多，主要有报国寺、万年寺、清音谷、金

- 峨眉山万佛顶（图片提供：全景正片）
Ten-thousand-buddha Summit of Mt. Emei

> Zhuyeqing (Green Bamboo Leaf) Tea from Mt. Emei

Mt. Emei is located in the southwest of Emei City of Sichuan Province. It is one of the four Buddhist holy lands in China. Mt. Emei has many famous historic and cultural sites, including the Baoguo (Serving the Motherland)

顶、白龙洞等。峨眉山报国寺的殿宇上方有一块匾额，上书"茶禅一味"四字，为中国佛教协会原会长赵朴初亲笔题词，它揭示了峨眉山茶与佛教结缘的深层关系。

峨眉山山区云雾多，日照少，随海拔高度的不同而呈现不同的气

Temple, Wannian (Ten Millennium) Temple, Qingyin (Clear Sound) Valley, Golden Summit, and Bailong (White Dragon) Cave. The board above the gate of the Baoguo Temple bears four Chinese characters meaning Tea and Zen in the Same Taste. Written personally by Zhao Puchu, a late president of the Buddhist

茶禅一味

佛教在中国兴起以后，由于坐禅需要而与茶结下了不解之缘，同时也为茶文化的传播作出了重要贡献，其核心是"茶禅一味"的理念。所谓"茶禅一味"就是说茶道精神与禅学相通、相近。僧人不只是饮茶止睡，而是通过饮茶意境的创造，将禅的哲学与茶结合起来。中国的"茶道"二字就是由禅僧首先提出的。

Tea and Zen in the Same Principle

After its rise in China, Buddhism has formed an indissoluble bond with tea due to the need of meditation. The development of Buddhism has also made important contribution to the spread of tea culture. The core of the bond is the concept of "Tea and Zen in the Same Principle". Such concept refers to the connection and approximation of teaism and Zen. Monks not only drink tea to stop sleep, but combine Zen philosophy with tea through the creation of a tea drinking atmosphere and prospect. The Chinese term for teaism was first put forward by Zen monks.

- "茶禅一味"匾额
  The Board of Tea and Zen in the Same Principle

候特征。山林茂密，物种多样，茶树生长环境良好，茶园主要分布在清音阁、白龙洞、万年寺、黑水寺一带，海拔均在800米左右。

峨眉山种茶，始于晋代。唐朝时其白芽茶就被列为贡茶，南宋诗人陆游曾将峨眉茶与名茶顾渚紫笋相提并论，认为这两种茶如同春兰秋菊，各有所长。1964年4月，陈毅副总理途经四川，来到峨眉山万年寺憩息。万年寺方丈用新采制的绿茶奉送给陈毅副总理品尝。陈毅副总理饮后顿觉清香沁脾，心旷神

- 竹叶青
  Zhuyeqing Tea

Association of China, these characters reveal the in-depth relationship between the tea produced in Mt. Emei and Buddhism.

Mt. Emei is covered with clouds and mists in most of the time and sunshine is rare here. The areas on different altitudes show different climate features. The mountain has lush forests, diversified species, and a good environment for the growth of tea trees. The tea gardens in the mountain are mainly distributed around the Qingyin Pavilion, Bailong Cave, Wannian Temple, and Heishui (Black Water) Temple. All these areas stand about 800 meters above sea level.

Tea planting in Mt. Emei started in the Jin Dynasty (265-420). In the Tang Dynasty (618-907), the Baiya (White Bud) Tea produced in the mountain was chosen as a tribute tea. Lu You, a famous poet in the Southern Song Dynasty (1127-1279), once equated the Emei Tea with the famous Zisun (Purple Bamboo Shoot) Tea from Mt. Guzhu and likened them to spring orchid and autumn chrysanthemum with their own merits. In April 1964, vice premier Chen Yi arrived in Sichuan Province on a business trip and took a rest at the Wannian Temple in Mt. Emei. Abbot of the temple presented

怡，疲劳顿消。在得知此茶尚未取名时，见此茶形、色似竹叶，且清香宜人，便为之取名"竹叶青"。

竹叶青茶鲜叶一般在清明前3～5天开采，标准为一芽一叶或一芽二叶初展，芽叶嫩匀，大小一致。竹叶青成品茶外形扁平光滑，两头尖细，形似竹叶，颜色翠绿，冲泡后汤色青澄，滋味浓醇，叶底嫩绿均匀。

- 竹叶青茶园
  A Zhuyeqing Tea Garden

the newly-made green tea to him. The tea was a good refreshment and vice premier Chen Yi liked it very much. When he was told that the tea was yet to be named, he called it Zhuyeqing because the tea had the shape and color of a bamboo leaf (*Zhu*: bamboo; *Ye*: leaf; *Qing*: green).

The fresh leaves of the Zhuyeqing Tea are normally picked three to five days before Pure Brightness. The picking is done when the first bud and first leaf or the first bud and first two leaves begin to open. The buds and leaves should be tender and even and in the same size. The finished Zhuyeqing Tea has a flat and smooth appearance. Its two ends are green and tapering like a bamboo leaf. The tea soup is green and clear with a rich flavor. The brewed tea leaves are tender, green, and even.

> 蒙山蒙顶茶

蒙山位于四川省雅安市境内，是著名旅游胜地，也是世界茶文化的发源地和人工栽培茶树最早的地区。唐宋以来，佛、道两教相继在蒙山建寺设观，山上现有千佛寺、净居庵、智矩寺、永兴寺、天盖寺等。山有五峰，即上清峰、菱角峰、毗罗峰、井泉峰和甘露峰，

- 蒙山茶园
  A Tea Garden in Mt. Mengshan

> Mengding (Mt. Mengshan Summit) Tea from Mt. Mengshan

Mt. Mengshan stands in the territory of Ya'an City of Sichuan Province. It is a famous tourist resort. It is also the birthplace of world tea culture and the first place for artificial planting of tea trees. Since the Tang and Song dynasties, many Buddhist temples and Taoist monasteries have been built in Mt. Mengshan, including the Qianfo (One-thousand-buddha) Temple, Jingju (Clean Residence) Temple, Zhiju (Wisdom Ruler) Temple, Yongxing (Forever Thriving) Temple, and Tiangai (Sky Dome) Temple. There are five peaks in the mountain, namely Shangqing (Upper Clear) Peak, Lingjiao (Water Chestnut) Peak, Piluo Peak, Jingquan (Well Spring) Peak, and Ganlu (Sweet Dew) Peak. They are like five petals of lotus in full

呈五瓣莲花盛开状。蒙山终年细雨蒙蒙，云雾茫茫，加上肥沃的土壤，为茶树的生长创造了适宜的条件。山上不但种茶，还有古时的皇家茶园。

由于蒙山茶主要产在山顶，故而所产名茶总称"蒙顶茶"。蒙山蒙顶茶自唐开始作为贡茶，一直延续到清代，长达1000余年，岁岁成为贡品，这在茶文化史上是罕见的。唐代时就有"蒙山顶上茶，扬子江心水"之说。蒙山顶上茶，当然指的是蒙顶茶；而扬子江心水，指的是被唐代品泉家刘伯刍评定为

● 蒙山皇家茶园
The Royal Tea Garden in Mt. Mengshan

blossom. Mt. Mengshan is covered with drizzle all the year round. Its cloudy and foggy weather and fertile soil provide tea trees with favorable growth conditions. In addition to ordinary tea gardens, there is an ancient royal tea garden in the mountain.

As the tea from Mt. Mengshan grows mainly on top of the mountain, it is called the Mengding (Mt. Mengshan Summit) Tea in general. The Mengding Tea from Mt. Mengshan served as a tribute tea for over 1,000 years from the Tang Dynasty to the Qing Dynasty. It was a rare case in tea culture history. In the Tang Dynasty (618-907), tea drinkers took two things seriously — the tea from the top of Mt. Mengshan and the water from the center of the Yangtze River. The tea from the top of Mt. Mengshan certainly referred to the Mengding Tea. The water from the center of the Yangtze River referred to the water of the Zhongling Spring in Zhenjiang City of Jiangsu Province, which was evaluated as the No.1 Spring in the World by Liu Bochu, a spring taster in the Tang Dynasty (618-907). Tea drinkers in the Tang Dynasty (618-907) put these two things together because they believed that the best tea should be brewed with the best water. A poet in the

"天下第一泉"的江苏镇江的中泠泉水。这句话说的是如此好茶，只有用天下最美的泉水才能相配。宋代时曾有诗人写诗赞誉蒙山顶上的茶胜过最好的美酒。

蒙顶茶中最著名的是蒙顶甘露和蒙顶黄芽。蒙顶甘露属炒青绿茶，外形紧卷多毫，嫩绿色润，香气馥郁，芬芳鲜清，汤色清澈微黄，叶底匀嫩秀丽。蒙顶黄芽为黄茶类茶中珍品，外形扁直，

Song Dynasty (960-1279) once wrote a poem to praise the tea from the top of Mt. Mengshan as a drink better than the best wine.

Among many Mengding teas, the most famous are Mengding Ganlu (Sweet Dew) and Mengding Huangya (Yellow Bud). The Mengding Ganlu is a stir fixation green tea. It is twisted and tippy and has a tender, green, and moist appearance and a strong and fresh aroma. The tea soup is clear with a yellowish tint. The brewed tea leaves are even, tender, and elegant. The Mengding Huangya is a top-grade product among yellow teas. It has a flat contour, a greenish yellow appearance, and apparent buds

• 冲泡后的蒙顶甘露
Brewed Mengding Ganlu

• 蒙顶甘露
Mengding Ganlu

色泽绿中显黄,芽毫显露,冲泡后甜香浓郁,汤色黄亮,滋味甘醇,叶底嫩黄、匀整,是蒙山茶中的极品。

and tips. The tea soup is sweet, fragrant, and brightly yellow, and tastes sweet and mellow. The brewed tea leaves are tenderly yellow and even. It is indeed the best of all the teas from Mt. Mengshan.

• 蒙顶黄芽
Mengding Huangya

黄茶

黄茶属于轻发酵茶,品质特点为黄叶黄汤、香气清悦、味厚爽口,主产于浙江、四川、安徽、湖南、广东、湖北等省。

黄茶与绿茶制作工艺相似,区别在于揉捻前或揉捻后,或在初干前或初干后进行焖黄。焖黄是黄茶加工的特点,是形成黄茶"黄叶黄汤"品质的关键工序。

黄茶根据鲜叶原料的嫩度和大小分为黄芽茶、黄小茶和黄大茶三类。黄芽茶,是以单芽或一芽一叶初展鲜叶为原料制成的,其品质特点是单芽挺直,冲泡后芽尖均朝上,直立悬浮于杯中,主要品种包括君山银针、蒙顶黄芽、莫干

• 温州黄汤(黄小茶)
Wenzhou Huangtang (Yellow Soup) Tea [Huangxiao (Yellow Small) Tea]

黄芽等；黄小茶又称"芽茶"，以细嫩的一芽一叶和一芽二叶初展制成，其品质特点是条索细紧显毫，汤色杏黄明净，滋味醇厚回爽，叶底嫩黄明亮，主要品种有沩山毛尖、北港毛尖、远安鹿苑、温州黄汤等；黄大茶又称叶茶，以一芽二三叶至四五叶为原料制成，其品质特点是叶肥梗粗，梗叶相连，金黄色，闻之轻发锅巴味，味浓耐泡，主要产于安徽霍山、金寨、六安、岳西和黄山，其品种有霍山黄大芽、广东大叶青等。

Yellow Tea

The yellow tea is a slightly-fermented tea featuring yellow leaves, yellow tea soup, clear and pleasant aroma, and strong and brisk taste. It is mainly produced in Zhejiang, Sichuan, Anhui, Hunan, Guangdong, and Hubei provinces.

The yellow tea is produced in the similar process for producing green tea. The difference is the step of yellow simmering before or after rolling or before or after the first drying. Yellow simmering is the unique step in yellow tea processing. It is a key process to achieve yellow leaves and yellow soup of the yellow tea.

Based on the tenderness and size of the raw material fresh leaves, the yellow tea can be divided into three types: Huangya (Yellow Bud) Tea, Huangxiao (Yellow Small) Tea, and Huangda (Yellow Big) Tea. The Huangya Tea is made with single bud or the freshly opening one bud and one leaf as the raw material. Its features include straight separate buds, whose tips point upwards after brewing, and vertical suspension of tea leaves in tea soup. Main varieties of this tea include Yinzhen (Silver Tip) Tea from Mt. Junshan, Mengding Huangya, and Huangya Tea from Mt. Mogan. Also known as Bud Tea, the Huangxiao Tea is made with fine, tender, and freshly opening one bud and one leaf or one bud and two leaves. Its features include fine, tight, and tippy strips. The tea soup is clear and apricot yellow. It tastes mellow and brisk. The brewed tea leaves are tenderly yellow and bright. Main varieties of this tea include Weishan Maojian (Hairy Tip), Beigang Maojian, Yuan'an Luyuan, and Wenzhou Huangtang (Yellow Soup). Also known as Leaf Tea, the Huangda Tea is made with one bud and two or three leaves to four or five leaves. Its features include fat leaves, thick stems, connected stem and leaf, and golden color. It has a slight smell of rice crust. The tea has a strong flavor and lasts long in brewing. It is mainly produced in Huoshan County, Jinzhai County, Liu'an City, Yuexi County, and Mt. Huangshan of Anhui Province. Its main varieties include Huangdaya (Yellow Big Bud) Ten from Huoshan County and Dayeqing (Big-leaved Green) Tea from Guangdong Province.

- 广东大叶青（黄大茶）
Dayeqing from Guangdong (Big-leaved Green) [Huangda (Yellow Big) Tea]

> 君山银针茶

> Yinzhen (Silver Tip) Tea from Mt. Junshan

君山是湖南岳阳市洞庭湖中的一座小岛，又称"洞庭山"，原本是神山仙境的意思。传说舜帝二妃娥皇和女英居此，而二妃又分别称

Mt. Junshan is a small isle in the Dongting Lake of Yueyang City, Hunan Province. It is also known as Mt. Dongting, which originally meant the divine mountain

● 君山小岛（图片摄影：杨一九）
Junshan Isle

为"湘妃"和"君妃"，故君山又有湘山之称。也有人说是因为秦始皇南巡泊此，遂名"君山"。这里四面环水，风光独好，景点颇多，著名的有洞庭庙、湘妃墓、朗吟亭、传书亭、秋月亭、云梦亭、烟波亭、望湖亭等。

传说四千多年前舜帝南巡时，其二妃娥皇和女英在白鹤寺旁亲自种茶树三棵。以后，经历代繁衍种植，才有了如今名闻迩遐的君山银针茶。君山产茶，有史可查的是始于唐代，因茶叶满披茸毛，底色泛黄，冲泡后像黄色羽毛一样竖立起

- 君山银针茶样、茶汤
Sample and Soup of the Yinzhen Tea from Mt. Junshan

and fairyland. Legend had it that two concubines of Emperor Shun, Ehuang and Nvying, once lived here. They were called Lady Xiang and Lady Jun, respectively. Therefore, Mt. Junshan is also known as Mt. Xiangshan. Another legend attributes the name of the mountain to a visit by the First Emperor of Qin (*Jun*: emperor). Mt. Junshan is surrounded by water on four sides and has a beautiful landscape. Among its famous scenic spots are the Dongting Temple, Tomb of Lady Xiang, Pavilion of Poem Chanting, Pavilion of Letter Passing, Pavilion of Autumn Moon, Pavilion of Cloud Dream, Pavilion of Smoky Waves, and Pavilion of Lake Sightseeing.

According to the legend, when Emperor Shun paid a visit to the south over 4,000 years ago, his two concubines, Ehuang and Nvying, personally planted three tea trees beside the Baihe (White Crane) Temple. Through generations of propagation, the famous Yinzhen Tea of Mt. Junshan came into being. Tea has been produced in Mt. Junshan since the Tang Dynasty (618-907). The tea leaves are covered with hairs on a yellow background. After brewing, they stand up in the water like yellow plumes, hence its another name Huanglingmao

来，一度被称为"黄翎毛"。君山银针茶也始于唐代，相传文成公主出嫁西藏时就选带了君山茶。君山茶在清朝时被列为贡茶。

君山银针的采摘讲究"九不采"：雨天不采、露水芽不采、紫色芽不采、空心芽不采、开口芽不采、冻伤芽不采、虫伤芽不采、瘦弱芽不采、过长过短芽不采。

君山银针以色、香、味、形俱佳而著称。制作成的茶芽壮挺直，长短大小均匀，内呈橙黄色，外裹一层白毫，故有"金镶玉"之称，又因茶芽外形像银针，故名君山银针。冲泡后的茶叶全部冲向上面，继而徐徐下沉，三起三落。茶汤色泽杏黄明亮，饮之味醇干爽，茶香四溢，叶底黄亮匀齐。

(Yellow Plume) in history. The Yinzhen Tea from Mt. Junshan also started in the Tang Dynasty (618-907). According to the legend, when Princess Wen Cheng was married to the King of Tibet, she took with her the tea from Mt. Junshan. The Junshan Tea was chosen as a tribute tea in the Qing Dynasty (1616-1911).

The Yinzhen Tea from Mt. Junshan should not be picked in rainy days. The following buds are not be picked either: dew bud, purple bud, hollow bud, open bud, bud with frost damage, bud with bug damage, lean bud, and too long or too short bud.

The Yinzhen Tea from Mt. Junshan is known for its color, aroma, taste, and appearance. The tea bud is thick and straight with even length and size. With an orange interior, it is covered with a layer of white hairs. It is therefore known as "Jade with Gold Inlay". As the tea bud looks like a silver needle, it is formally called Junshan Yinzhen. After brewing, the tea leaves first float on the surface and then slowly sink. This process goes on for three rounds. The tea soup is apricot yellow and bright and tastes mellow and brisk. The tea aroma is profound and spreads far and wide. The brewed tea leaves are yellow, bright, and even.

> 六大茶山普洱茶

在美丽的云南西双版纳，有六处连片的山岭，合称为"六大茶山"，这便是普洱茶的原产地。普洱茶产区有古今六大茶山之说。古六大茶山是指攸乐、革登、倚邦、

- 西双版纳风光
Landscape of Xishuangbanna Dai Autonomous Prefecture

> Pu'er Tea from Six Major Tea Mountains

There are six integrated mountains in the beautiful Xishuangbanna Dai Autonomous Prefecture of Yunnan Province. Known as "the Six Major Tea Mountains", they are the place of origin of the Pu'er Tea. In the ancient times, people called the mountains on the northern bank of the Lancang River Six Major Tea Mountains, namely Youle, Gedeng, Yibang, Mangzhi, Manzhuan, and Mansa mountains. Now, people regard another six mountains on the southern bank of the Lancang River as current Six Major Tea Mountains. These mountains are Nannuo, Nanqiao, Mengsong, Jingmai, Bulang, and Bada. Now, the current Six Major Tea

莽枝、蛮砖、曼撒六座茶山，全都位于澜沧江北岸；今六大茶山是指南糯、南峤、勐宋、景迈、布朗、巴达六座茶山，全都位于澜沧江南岸。今六大茶山是近现代普洱茶的主要原料产地。

云南是茶的发源地之一，早在3000多年前就已经开始有人工种植茶树。到唐朝时，普洱茶才开始大规模地种植和生产，那时称为"普茶"。清朝时，普洱茶达到鼎盛，并被列为贡茶，也常常作为国礼赠

• 普洱茶茶样、茶汤
Sample and Soup of the Pu'er Tea

• 古六大茶山分布示意图
古六大茶山山水相连，山清水秀，云雾缭绕，气候温暖湿润，日照充足，且土壤多为微酸性，十分适合茶树生长。
Schematic Diagram of the Distribution of the Ancient Six Major Tea Mountains
The ancient Six Major Tea Mountains stand side by side with each other and form a beautiful landscape. They enjoy sufficient sunshine and a warm and humid climate. Their soil shows slight acidity. Always covered with cloud and mist, they have an ideal condition for the growth of tea trees.

Mountains are the main producers of the raw materials for making modern Pu'er Tea.

Yunnan Province is one of the birthplaces of tea. People began planting tea trees in Yunnan Province over 3,000 years ago. By the Tang Dynasty (618-907), the Pu'er Tea had been under large-scale planting and production. It was called Pu Tea at that time. In the Qing Dynasty (1616-1911), the Pu'er Tea reached its zenith and was chosen as a tribute tea and a national gift for foreign envoys. In recent years, the Pu'er Tea is loved by people due to its high efficacy in health care and health preserving. A new heyday has come for the Pu'er Tea.

送给外国使节。近年来，具备保健、养生功效的普洱茶更是得到人们的追捧，再次进入鼎盛时期。

普洱茶最根本的品质特征在于耐久。茶叶一般来说以新鲜为好，但普洱茶可以在自然中呼吸，在空气中持续发酵，存放越久茶香越醇。普洱茶的香气高锐持久，带有云南大叶种茶的独特香型，滋味浓强，富于刺激性；芽壮叶厚，白毫密布，经五六次冲泡后仍有香味，汤色橙黄浓厚。

The most fundamental feature of the Pu'er Tea is its endurance. Normally, fresh tea is better than the old one. However, the Pu'er Tea can breathe in a natural environment and keep on fermentation in the air. The longer it is stored, the mellower it becomes. The Pu'er Tea has a strong and long-lasting aroma, the one peculiar to Dayezhong (Big leave) Tea. Its taste is strong. It has sturdy buds and thick leaves covered with white hairs. After five or six rounds of brewing, its aroma still remains. The tea soup is orange and thick.

- 澜沧江畔的巍巍茶山
  Towering Tea Mountains by the Lancang River

黑茶

黑茶，属后发酵茶，是中国特有的茶类。其所用原料较粗老，制造过程中堆积发酵时间较长，是一种成茶色泽油黑或黑褐色的茶种。其品质特征为色泽黑褐油润，汤色橙黄或橙红。主要产于湖南、湖北、广西、云南、四川等地。

黑茶生产历史悠久，其工艺流程包括杀青、揉捻、渥堆、干燥等工序，其中最重要的工序是渥堆。

黑茶的主要品种有云南的普洱茶、湖南的黑毛茶、湖北的老青茶、四川的南路边茶与西路边茶、广西的六堡茶等，其中以云南的普洱茶最负盛名。普洱茶属云南大叶种茶，每克云南大叶种茶的水浸出物、茶多酚、儿茶素均远高于中国其他名茶，因而具有更显著的抗衰老功效。另外，因为茶叶在发酵的过程中形成的多种有益菌群，可减少小肠对三酸甘油酯和糖的吸收，提高酵素分解脂肪，所以经过发酵的普洱熟茶在减肥方面也能起到更有效的作用。

- 湖南安化黑茶
  Dark Tea of Anhua City, Hunan Province

Dark Tea

The dark tea is a post-fermented tea, a unique tea variety of China. It is made with coarse and old raw materials through a long period of piled fermentation. The finished tea is oily black or blackish brown. The tea soup is orange or orange red. It is mainly produced in Hunan, Hubei, Guangxi, Yunnan, and Sichuan.

The dark tea has been produced for a long time. Its processing steps include fixation, rolling, piling, and drying. Among them, piling is the most important step.

- 普洱茶
  Pu'er Tea

Among the major dark teas are the Pu'er Tea of Yunnan Province, Heimao (Black Hair) Tea of Hunan Province, Laoqing (Old Black) Tea of Hubei Province, Nanlubian (Southern Roadside) Tea and Xilubian (Western Roadside) Tea of Sichuan Province, and Liubao (Six-castle) Tea of Guangxi Zhuang Autonomous Region. Among them, the Pu'er Tea of Yunnan Province is the most famous. It is a kind of Yunnan Dayezhong Tea, which contains far greater aquatic extract, GTP, and catechin than other famous teas of China. It therefore shows greater efficacy in resisting aging. In addition, several kinds of helpful bacterial groups will grow during fermentation. These bacterial groups can reduce the absorption of triacid glyceride and sugar by small intestine and boost fat decomposition by ferment. Therefore, the fermented Pu'er Tea can also play an effective role in reducing weight.

- 渥堆发酵的普洱茶

渥堆是指将晒青毛茶堆放起来后洒水，置于室温25℃以上，相对湿度85%左右的环境内，促使其发酵的过程。这是加速普洱茶陈化的一种工序，也是形成和奠定普洱茶特殊品质的关键工艺。

Pu'er Tea under Pile-fermentation

Piling refers to the process during which the sunned semi-made tea is piled up, watered, and stored in an environment with a temperature of over 25℃ and a relative humidity of 85% to promote its fermentation. Piling is for speeding up the aging of the Pu'er Tea. It is a key step to form and consolidate the special quality of the Pu'er Tea.

普洱茶的分类

　　根据制法的不同，普洱茶可以分为生茶和熟茶两种。生茶是指采摘以后自然干燥而成，茶饼多为青绿色、墨绿色，少部分为黄红色，茶性较烈，存放数年后会渐渐变得温和，冲泡出的汤色一般为青绿色，收藏用的普洱茶大多是生茶；熟茶是指以人工渥堆发酵制成的茶，茶饼多为黑色或红褐色，部分茶芽成暗金黄色，多有渥堆味，茶性较温和，冲泡后的茶汤色泽金红。

　　按照存放方式的不同，普洱茶可以分为干仓普洱和湿仓普洱。干仓普洱是指存放于通风干燥的仓库，自然发酵而成的普洱茶，这种茶较完整地保留了普洱茶的味道，越陈越香，具有很高的收藏价值；湿仓普洱茶是指存放于较潮湿的地方，加速发酵而成的普洱茶，但这种茶因陈化速度快而易产生霉变，对人体不利。

　　按照外形不同，普洱茶可以分为饼茶、沱茶、砖茶、金瓜茶、条茶、散茶等。

Classification of the Pu'er Tea

According to the method of production, the Pu'er Tea can be divided into two kinds: unfermented and fermented. The unfermented tea is naturally dried after picking and its caky tea is mostly green and dark green, and a few are yellowish red. The tea tastes strong. After several years' storage, it will turn mild and produce green tea soup. Most of the Pu'er Tea for collection is unfermented. The fermented tea has gone through pile-fermentation and its caky tea is mostly black or reddish brown. Some tea buds are dark golden. Most fermented Pu'er Tea has the piling smell and tastes mild. The tea soup is golden red.

　　According to the storage modes, the Pu'er Tea can be divided into dry-warehouse Pu'er Tea and wet-warehouse Pu'er Tea. Dry-warehouse Pu'er Tea is made through natural fermentation in a dry and well-ventilated warehouse. It retains the taste of the Pu'er Tea rather completely and has a high collection value since it acquires stronger aroma through storage. Wet-warehouse Pu'er Tea is made through speed-up fermentation in a damp warehouse. This tea will easily become moldy due to speedy aging process, which is harmful to drinker's health.

　　According to the appearance, the Pu'er Tea can be divided into the following types: caky tea, bowl-shaped tea, brick-shaped tea, melon-shaped tea, bar tea, and bulk tea.

- 生茶茶饼
Unfermented Caky Tea

- 熟茶茶饼
Fermented Caky Tea

- 普洱饼茶

饼茶，又称"七子饼茶"，是指呈扁平圆盘状的普洱茶。每块七子饼茶重七两（375克），每七个为一筒，七七四十九，代表多子多孙的含义。

Caky Pu'er Tea

The caky tea is also known as the Seven-son Caky Tea. Shaped like a flat round disc, each piece of the Seven-son Caky Tea weighs 375 grams. Seven tea cakes are packed in one box. Seven times seven is forty-nine, representing flourishing posterity.

- 普洱沱茶

沱茶，是指碗状的蒸压茶，一般每个净重100克或250克。近年来也生产了一些迷你的小沱茶，平均每个净重2～5克。

Bowl-shaped Pu'er Tea

The bowl-shaped tea is made through steam-pressing. Normally, each piece weighs 100 grams or 250 grams. In recent years, some mini bowl-shaped teas are produced and each piece weighs two to five grams.

- 金瓜茶

金瓜贡茶，是指压制成大小不等的瓜形的普洱茶，每块100克到数百千克均有。

Melon-shaped Pu'er Tea

The Pu'er Tea can be compressed into melon shapes in different sizes. The weight of each piece ranges from 100 grams to several hundred kilograms.

- 普洱散茶

散茶，是指没有经过紧压的散装普洱茶。

Bulk Pu'er Tea

There is also the uncompressed bulk Pu'er Tea.

- 普洱砖茶

砖茶，是指长方体或正方体形状的普洱紧压茶，每块250～1000克。

Brick-shaped Pu'er Tea

Brick-shaped Pu'er Tea is compressed into its rectangular solid or cube shape and each piece weighs 250 to 1,000 grams.

名山出好茶

Good Teas from Great Mountains

> 太姥山绿雪芽茶

　　太姥山，旧称"太母山"，又称"才山"，位于福建省福鼎市境内东北部。太姥山海拔917米，周围20千米，以"峰、石、洞、雾"四绝称雄，共有54峰、45石、24洞、9泉、3溪，景点甚多。这里群山叠

• 太姥山风光
Landscape of Mt. Taimu

> Lvxueya (Green Snow Bud) Tea from Mt. Taimu

Also called Mt. Caishan (Talent Mountain), Mt. Taimu is located in the northeast of Fuding City of Fujian Province. It stands 917 meters above sea level and is 20 kilometers around. Known for its beautiful peaks, rocks, caves, and fog, the mountain has 54 peaks, 45 rocks, 24 caves, nine springs, three streams, and numerous scenic spots. Here, green mountains extend as far as the eye can see and rocky canyons are contending in beauty and fascination. According to the legend, Taoist Rongchengzi once practiced alchemy here. The well from which he collected water for pill making still remains.

　　Lvxueya Ten is called the Divine Tea. It originally grew beside the Hongxue (Great Snow) Cave in Mt. Taimu in the south of Fuding City, Fujian Province.

太姥山鸿雪洞旁的绿雪芽茶树
The Lvxueya Tea Tree by the Hongxue Cave in Mt. Taimu

翠，岩壑争奇。相传早在黄帝时，就有道家仙人容成子在此炼丹，当年炼丹用水的炼丹井，至今犹存。

　　绿雪芽，人称"仙茶"，早年生长于福建福鼎市南太姥山鸿雪洞旁。相传尧帝时，山下有一老妪为避乱上山，以种兰为业。一天，老妪路过鸿雪洞，发现一株茶树，当即培土，并用炼丹井水灌浇，取名"绿雪芽"，这就是当今名茶福鼎大白茶的祖先。据说这种茶治疗小儿麻疹有特效。有一年山下村中麻疹大流行，村民无药可治，老妪便用绿雪芽治好不少病孩。为此，村民感激不已，尊称老妪为"太母

The story goes that when Emperor Yao was in throne, an old woman went into the mountain in pursuit of a peaceful life. She made a living by planting orchids. One day, she found a tea tree beside the Hongxue Cave. She immediately earthed up the tea tree and watered it. She named the tea tree Lvxueya (Green Snow Bud), which was the ancestor of Dabai (Big White) Tea from Fuding City, a present-day famous tea. It is said that the tea has special effect in curing measles in children. One year, measles broke out in a village at the foot of Mt. Taimu and the villagers could do nothing about it. The old woman cured many children with Lvxueya Tea and the villagers were thankful and respectfully called her Goddess Taimu. After she died, the villagers buried her in the mountain and called the mountain Mt. Taimu. In the Han Dynasty (206 BC-220), Emperor Wudi ordered Dongfang Shuo, one of his trusted officials, to entitle Mt. Taimu a famous mountain.

　　White-tipped Yinzhen Tea from Fuding City is a white tea made with the fresh leaves, one bud and one leaf, from the improved Dabai Tea tree. The finished tea looks like silver white needles covered with white hairs. The

娘娘"。老妪死后，葬于此山，村民也就称其为"太母山（太姥山）"。汉代，重臣东方朔奉汉武帝之命，敕封太姥山为天下名山。

福鼎白毫银针是以大白茶良种茶树鲜叶为原料制成的白茶。其采制标准为一芽一叶，成茶形状似针，白毫密被，色白如银。冲泡后汤色浅黄清澈，香气清鲜，滋味醇厚爽口，叶底嫩匀完整，也是茶中佼佼者。

太姥山白琳、湖林一带还产有白琳工夫红茶。这里山清水秀，终年云遮雾绕，很适宜茶树生长。生长在太姥山区的茶树，根深叶茂，芽毫雪白晶莹，用这种茶树芽叶制成的工夫红茶品质上佳。白琳工夫红茶最初采用小叶种茶树鲜叶为原料制成，20世纪初改用福鼎大白茶良种为原料。其干茶色泽黄黑，条索细长弯曲，茸毫多呈颗粒绒球状，冲泡后汤色浅红明亮，香气鲜嫩纯正，滋味清鲜回甜，叶底鲜红带黄。

tea soup is light yellow and clear. With a fresh aroma, it tastes mellow and brisk. The brewed tea leaves are tender, even, and complete. It is also one of the best teas.

The areas around Bailin and Hulin in Mt. Taimu also produce Congou Black Tea. The areas are picturesque and covered with cloud and fog all the year round, which are good for the growth of tea trees. The tea trees in Mt. Taimu have deep roots and lush leaves. Their bud tips are white and crystal-clear like snow. The Congou Black Tea made with the buds from such tea trees has the best quality. The Bailin Congou Black Tea was initially made with the fresh leaves from small-leaved tea trees. By the beginning of the 20th century, it was made with the improved Dabai Tea from Fuding City. The dry tea leaves are yellowish black and shape like long, thin, and bent strips. The tea hairs look like particle pompons. The tea soup is light red and bright, with a fresh, tender, and pure aroma. It tastes fresh and sweet. The brewed tea leaves are bright red with a yellow tint.

• 白毫银针茶样、茶汤
Sample and Soup of White-tipped Yinzhen Tea

白茶

　　白茶是一种轻微发酵茶，因其表面披满白色茸毛而得名。其主要产于福建省的福鼎、政和、松溪、建阳等地。

　　白茶的品质特点是成茶外表披满白毫，绿叶红筋。茶叶冲泡后叶片完整而舒展，茸毛多，汤色浅淡或初泡无色，香味醇和。

　　白茶是加工方式最简单的茶，有"懒人做茶做白茶"之说，其制作工艺一般只有萎凋和干燥两道工序，其中萎凋是关键。但加工时对鲜叶的要求非常严格，要求具备"三白"，即嫩芽和两片嫩叶披满白色茸毛。

　　白茶的主要品种有白毫银针、白牡丹、贡眉等。

White Tea

The white tea is a slightly-fermented tea. It derives its name from the white hairs on its surface. It is mainly produced in Fuding City, Zhenghe County, Songxi County, and Jianyang District of Fujian Province.

　　The finished white tea has green leaves and red veins covered with white hairs. The brewed tea leaves are complete. They fully open up in the water and are covered with many hairs. The tea soup has a light color and a mellow aroma. Sometimes, the first brewing shows no color.

　　The white tea requires the simplest processing procedure and is often referred to as the tea made by lazy producers. It goes through only two procedures: withering and drying. Of them, withering is the key. It has strict three-white requirements on fresh leaves: the tender bud and two tender leaves should be covered with white hairs.

　　Major white tea varieties include White-tipped Yinzhen Tea, Baimudan (White Peony) Tea, and Gongmei Tea.

- 白毫银针
White-tipped Yinzhen Tea

- 白牡丹
Baimudan (White Peony) Tea

- 贡眉
Gongmei Tea

红茶

　　红茶是完全发酵的茶类,最早源于福建省崇安地区(今武夷山市),距今有200多年的历史。其品质特点是红汤红叶。

　　红茶产区主要集中于华南茶区的海南、广东、广西、台湾以及湖南和福建南部;西南茶区的云南、四川等地;江南茶区的安徽、浙江、江西也有少量生产。

　　中国生产的红茶主要有小种红茶、工夫红茶和红碎茶三个类别。

　　红茶的生产工艺要经过萎凋、揉捻、发酵(渥红)和干燥等工序。另外,小种红茶在制作中还要增加过红锅和薰焙两个工序。

　　中国红茶中的名品有祁门红茶、滇红、宁红等。

● **祁门红茶茶样、茶汤**

祁门红茶,因原产于安徽省祁门县而得名,是工夫红茶中的极品,多年来一直是中国的国事礼茶。它与印度的大吉岭红茶、斯里兰卡的乌伐红茶并称为世界三大高香红茶。

Sample and Soup of the Keemun Black Tea

The Keemun Black Tea derives its name from its place of origin, Keemun (Qimen) County of Anhui Province. As a top product among Chinese congon black teas, it has long been serving as a national gift tea of China for foreign envoys. It is one of the top three high-aroma black teas in the world. The other two are the Darjeeling Black Tea of India and Uva Black Tea of Sri Lanka.

Black Tea

A fully-fermented tea, the black tea originated in Chong'an District (present-day Wuyishan City) of Fujian Province over 200 years ago. Its features include red tea soup and blackish red leaves.

The black tea is produced mainly in three areas: South China Area, Southwest Area, and the Area South of the Yangtze River. The South China Area where the black tea is mainly produced includes Hainan, Guangdong, Guangxi, Taiwan, and south of Hunan and Fujian. The Southwest Area where the black tea is mainly produced includes Yunnan and Sichuan. The Area South of the Yangtze River where the black tea is mainly produced includes Anhui, Zhejiang, and Jiangxi, where small quantity of the black tea is produced.

The black tea produced in China includes three varieties: Souchong Black Tea, Congou Black Tea, and Broken Black Tea.

The black tea is made through withering, rolling, fermentation, and drying. For Souchong Black Tea, two more procedures are needed: hot pot frying and steam baking.

Among the famous black teas made in China are the Keemun Black Tea, Dianhong (Yunnan) Black Tea, and Ninghong Black Tea.

• 红碎茶
Broken Black Tea

• 小种红茶
Souchong Black Tea

> 武夷山大红袍

武夷山，位于福建省武夷山市境内，风光秀丽，是著名的旅游胜地，著名的武夷九曲溪盘绕于群山之间。武夷山出名，与其盛产的武夷岩茶也有很大的关系。

• 武夷山风光
Landscape of Mt. Wuyi

> Dahongpao (Big Red Robe) Tea from Mt. Wuyi

Mt. Wuyi stands in the territory of Wuyishan City of Fujian Province. With beautiful landscape, it is a famous tourist resort. The well-known Wuyi Jiuqu (Nine-turn) Stream flows through the mountain. In fact, the Wuyi Rock Tea has played an important role in putting Mt. Wuyi on the map.

Mt. Wuyi has precipitous cliffs, deep pits, and huge valleys. The ancient people planted tea trees in the rock pits, crevices, and cracks by building embankment with rocks. Year in year out, Mt. Wuyi has ended up with rocks in every corner and tea trees amidst all rocks. People call the oolong tea produced in Mt. Wuyi the Wuyi Rock Tea. There are many types of the Wuyi Rock Tea. For example, Qizhong Tea is none but the Vegetable Tea as called by the locals, because it

武夷山悬崖绝壁，深坑巨谷。先人利用岩凹、石隙、石缝，砌石为岸，从中栽茶。长年累月，形成武夷山"山山有岩，岩岩有茶"的格局。人们将产于武夷山的乌龙茶称为"武夷岩茶"。武夷岩茶种类繁多，例如，奇种即当地所谓的"菜茶"，得名于它们普通得跟日常吃的菜一样；单丛奇种是从菜茶中选育而成，再冠以各种名称；名丛奇种是单丛奇种中最优秀的，最著名的有"四大名丛"：大红袍、水金龟、白鸡冠、铁罗汉。武夷岩茶驰名海外与名丛多分不开。名丛是"岩茶之王"，而在四大名丛中，尤以大红袍为贵，称为武夷岩茶中的"王中之王"。

is as plain as the ordinary vegetable. Dancong Qizhong Tea is developed from the Vegetable Tea and then given various names. Mingcong Qizhong Tea is the best of Dancong Qizhong Tea. There are top four Mingcong Qizhong Tea: Dahongpao (Big Red Robe), Shuijingui (Water Golden Turtle), Baijiguan (White Comb), and Tieluohan (Iron Arhat). The Wuyi Rock Tea is well-known at home and abroad, which has much to do with Mingcong Qizhong Tea, the kings of rock tea. Among the top four Mingcongs, Dahongpao Tea is the best and it is called the king of kings of the Wuyi Rock Tea.

There are only six Dahongpao tea trees. They grow on the rock wall of Jiulongke (Nine-dragon Den), Tianxin (Sky Heart) Rock, Mt. Wuyi. The site

- 白鸡冠茶样、茶汤

Sample and Soup of the Baijiguan (White Comb) Tea

- 水金龟茶样、茶汤

Sample and Soup of the Shuijingui (Water Golden Turtle) Tea

• 大红袍
Dahongpao Tea

• 铁罗汉茶样、茶汤
Sample and Soup of the Tieluohan Tea

    大红袍生长在武夷山天心岩九龙窠的岩壁上，仅有六棵茶树。这里日照时数短，气温变动大。岩壁上有细小甘泉，终年缓缓滴流，滋润着大红袍茶树的生长。从山岩不断垂落的枯枝落叶，落在大红袍茶树根际，又为大红袍茶树提供了丰厚的营养。如此天赋不凡、得天独厚的条件，造就了大红袍茶树的最佳生态环境。大红袍的采制至今约有300年的历史。由于大红袍珍贵奇异，所以古时采制时，需焚香礼拜，设坛诵经，使用特制的器具，挑选精练的制茶高手，方能制得。

    大红袍品质很有特色，成茶条紧，色泽绿褐，冲泡后的汤色橙黄，香气馥郁，具有持久的桂花香，这种

enjoys little sunshine and is subject to large temperature difference each day. A small spring flows through the rock wall all the year round and provides water to the tea trees. The withered branches and leaves from above the rock fall to the root of the tea trees, offering rich nutrition to them. Such ideal natural conditions contribute to the high quality of the Dahongpao Tea. Dahongpao Tea has been made for 300 years. It was so highly valued in the past that only the master tea makers were chosen to pick and make the tea. In the process, the master tea makers burnt incense, set up an altar to offer sacrifice and chant classics, and only used the utensils dedicated to the making of the Dahongpao Tea.

    Dahongpao Tea has a unique

韵味只有大红袍特有。乌龙茶通常能冲泡三四次；名丛能冲泡五六次，好的能做到"七泡有余香"；而大红袍冲至八九次，仍不脱原茶真味桂花香，不愧为"茶中之王"。

- **武夷大红袍**

 "大红袍"三个字的石刻是1927年由天心寺和尚所做，至今仍在，旁边的六株大红袍母树都属灌木茶丛。

Dahongpao Tea from Mt. Wuyi

The stone carving of the three characters, Da Hong Pao, was made by monks of the Tianxin (Sky Heart) Temple in 1927. The six parent Dahongpao tea trees beside it are shrub-type tea trees.

quality. The finished tea is seen as tight and greenish brown strips. The orange tea soup sends forth strong aroma, like the lasting fragrance of sweet-scented osmanthus. Such taste is peculiar to Dahongpao Tea. The oolong tea can last three to four rounds of brewing. The Mingcong Tea can last five to six rounds. The high-quality one still has aroma even after seven rounds of brewing. Dahongpao Tea, can last eight to nine rounds of brewing and still keep its fragrance. It is indeed the King of Teas.

乌龙茶

乌龙茶，即青茶，属半发酵茶，是中国独具特色的茶叶品类，主要盛产于福建、广东和台湾三地。乌龙茶既有红茶的浓、鲜、醇，又有绿茶的清香，但没有绿茶的苦味和红茶的涩味。

乌龙茶的生产始于16世纪末17世纪初，加工工艺很特别：鲜叶需经晒青、凉青、做青、炒青（相当于绿茶杀青）、揉捻和烘焙等工艺制成。干茶色泽通常为青褐色，茶汤黄亮，叶底明亮，滋味醇厚耐泡。乌龙茶冲泡后，叶片展开有绿有红，叶中间呈绿色，叶缘呈红色，因此有"绿叶红镶边"的美称。

乌龙茶的养生保健功能在茶叶中属佼佼者。它除了具备一般茶叶的保健功能外，还具有抗衰老、抗癌症、抗动脉硬化、防治糖尿病、减肥健美、防止龋齿、清热降火、敌烟醒酒等功效。

乌龙茶中的名品有武夷山大红袍、安溪铁观音、凤凰山单丛茶、台湾冻顶乌龙茶等。

- 台湾顶级乌龙茶的干茶、茶汤、叶底

Dry tea, Soup, and Brewed Tea Leaves of the Top-grade Oolong Tea from Taiwan Province

- 乌龙茶

Oolong Tea

Oolong Tea

The oolong tea is a semi-fermented tea, a unique tea variety of China. It is mainly produced in Fujian, Guangdong, and Taiwan Provinces. The oolong tea has the richness, freshness, and mellowness of black tea and the fragrance of green tea, but does not have the bitterness of green tea and the astringent taste of black tea.

Production of oolong tea started by the end of the 16th century or the beginning of the 17th century. Its processing technique is very special. The fresh leaves should go through sunning, cooling, stir fixation (equivalent to green tea fixation), rolling, and baking. The dry tea is greenish brown and the tea soup is yellow and bright. The brewed tea leaves are bright. The tea soup tastes mellow and remains so after repeated brewing. After brewing, the tea leaves open up. The middle part is green and the edge is red, hence its nickname as "Green Leaves with Red Edge".

The oolong tea ranks first in preserving health. In addition, it performs well in resisting aging, cancer, and arteriosclerosis, and preventing and curing diabetes, reducing weight and keeping fit, preventing tooth decay, clearing heat, and countering the influence by cigarette smoking and alcohol.

Among the famous oolong teas are the Dahongpao from Mt. Wuyi, Tieguanyin Tea from Anxi County, Dancong Tea from Mt. Phoenix, and Dongding Oolong Tea from Taiwan Province.

安溪铁观音

安溪铁观音，又名"红心观音""红样观音"，是用铁观音茶树鲜叶为原料制成的乌龙茶，产于福建省安溪县。这里溪水清澈长流，高山云雾缭绕，气候湿润温暖，雨水丰富，非常适宜乌龙茶树的生长。

安溪种茶始于唐末，五代时逐渐发展，并开始将茶作为礼品赠送。宋代时，寺观和农家均有产茶。明代时，安溪茶农发明了茶树无性繁殖法，即茶树整株压条繁殖法。在此之前，中国茶树均采用种子直播的有性繁殖法，茶树容易变种。因此，安溪是中国茶树无性繁殖的发源地。明末清初，安溪茶农创制了乌龙茶，是对传统制茶工艺的一次重要革新。后来，安溪茶农又发现了名茶铁观音，这一发现奠定了安溪作为中国名茶之乡的地位。

安溪铁观音一年可采四至五季，即春茶、夏茶、暑茶、秋茶和冬片。品质以春茶最好，秋茶次之。茶叶的品质还与采摘时的天气有关。晴天采摘的最好，阴天次之，雨天最差。

• 铁观音
Tieguanyin Tea

安溪铁观音茶条卷曲，叶表带白霜，肥壮圆结，色泽砂绿，鲜润显红点；整体形状似青蒂绿腹蜻蜓头、螺旋体、青蛙腿。冲泡后汤色金黄浓艳似琥珀，有天然馥郁的兰花香，滋味醇厚甘鲜，回甘悠久，茶香高而持久，可谓"七泡有余香"。叶底肥厚明亮，具绸面光泽。

Tieguanyin Tea from Anxi County

Also known as Hongxin Guanyin and Hongyang Guanyin, Tieguanyin Tea from Anxi County is an oolong tea made with the fresh leaves of the Tieguanyin tea trees growing in Anxi County of Fujian Province. The area has clear springs flowing all the year round. The high mountains in the area are always covered with clouds and mists. The humid and warm climate and plenty of rainfall in the area are good for the growth of the oolong tea trees.

Tea planting in Anxi County started at the end of the Tang Dynasty and developed in the Five dynasties. Gradually, the tea produced in Anxi County was chosen as a gift. In the Song Dynasty, both temples and farmer households in Anxi planted tea. In the Ming Dynasty, Anxi tea farmers created tea tree vegetative propagation, or whole-tea-tree mound layering. Before that, tea trees in China were subject to generative propagation via direct seeding, which could easily lead to the change of tea tree varieties. Therefore, Anxi County is the birthplace of tea tree vegetative propagation in China. By the end of the Ming Dynasty and the beginning of the Qing Dynasty, Anxi tea farmers created the oolong tea, marking a major renovation on traditional tea making. Later, Anxi tea farmers discovered Tieguanyin, which later became a famous tea. This discovery laid the foundation for Anxi to become the hometown of famous teas of China.

Tieguanyin Tea from Anxi County can be picked four to five times a year and made into spring tea, summer tea, midsummer tea, autumn tea, and winter tea. The spring tea is the best while the autumn tea the second best. Tea quality also depends on the weather. The tea picked on sunny days is the best, followed by those picked on cloudy days and rainy days.

Tieguanyin Tea from Anxi County seen as curly strips covered with white hairs. The tea strips are fat and round with fresh and moist red spots on green background. It seems to have the green dragonfly head, spiral body, and frog leg. The tea soup is golden as amber and sends forth natural rich orchid fragrance. It tastes mellow, sweet, and fresh, with long-lasting sweet aftertaste. The tea has high aroma that lasts long, as long as after seven rounds of brewing. The brewed tea leaves are fat, thick, and bright, with the luster seen on the silk surface.

• 安溪铁观音茶园
A Tieguanyin Tea Garden in Anxi County

> 凤凰山单丛茶

凤凰山位于广东省潮州市北的潮安县境内，主峰高约1500米，为潮州地区的第一高峰。这里濒临南海，气候温暖，雨量充沛，终年云雾弥漫，空气湿润。这种独特的生态环境孕育了茶树良种——凤凰水仙。

> Dancong (Single Clump) Tea from Mt. Phoenix

Mt. Phoenix is located in Chao'an County to the north of Chaozhou City of Guangdong Province. About 1,500 meters above sea level, its main peak is the highest peak in Chaozhou City. The area is close to the South China Sea and

- 凤凰山
  Mt. Phoenix

凤凰山种茶，相传始于南宋末年，宋帝赵昺路经凤凰山区的乌岽山时，口渴难耐，侍从们采当地茶树嫩叶片，烹制成茶汤解渴。赵昺饮后，顿觉口舌生津，奇香无比，于是称赞乌岽山茶风韵奇特，最能解渴。从此，潮安民众广为种植，并称这种茶树为"宋种"。如今，在乌岽山的石壁上，仍刻有"宋种"两字，以示纪念。在凤凰山的乌岽山上，至今还保留着自宋起，历经元、明、清，直至现代的不同

enjoys a warm climate and plenty of rainfall. It is covered with cloud, mist, and moist air all the year round, making it an ideal place to cultivate Fenghuang (Phoenix) Shuixian (Narcissus) Tea, an improved tea tree variety.

According to the legend, tea planting in Mt. Phoenix started in the end of the Southern Song Dynasty. At that time, Emperor Zhao Bing passed by the Wudong Hill in Mt. Phoenix, and he was thirsty. His attendants picked some tender leaves from local tea trees and cooked the tea for him. The tea proved a great

• 凤凰乌岽山天池
The Tianchi Lake on Top of the Wudong Hill of Mt. Phoenix

生长树龄的茶树千余株，树龄最大的达800年，被称为宋种后代。

凤凰单丛是选用树型高大的凤凰水仙群体品种中的优异单株单独采制而成。其成茶挺直肥硕，色泽黄褐，有天然花香，且香气浓烈。冲泡后的茶汤滋味浓郁，甘醇爽口，有特殊的山韵蜜味，汤色清澈，耐冲泡。

凤凰单丛系列乌龙茶至少有80

● 凤凰单丛茶样
Sample of Phoenix Dancong Tea

● 凤凰单丛茶汤
Soup of Phoenix Dancong Tea

relief for Zhao Bing and he highly praised it. Since then, people in Chao'an widely planted tea trees and named local tea tree "Song Strain". Today, on the cliff of the Wudong Hill, the two characters "Song Strain" can still be seen. In the Wudong Hill of Mt. Phoenix, there are still over one thousand tea trees planted in different dynasties, including the Song, the Yuan, the Ming, the Qing dynasties and even the modern times. The oldest tea trees are 800 years old and they are the Offspring of Song Strain.

Dancong Tea form Mt. Phoenix is made with only the leaves of the high-quality individual tea trees among the tall Phoenix Narcissus group variety. The finished tea is straight, fat, and yellowish brown, with intense natural flower scent. The tea soup is clear and has a strong sweet and brisk taste and a special mountain taste and honey flavor, which linger even after repeated brewing.

There are at least over 80 varieties of Dancong oolong tea from Mt. Phoenix. The most famous ones include Yellow Branch Aroma Tea, Cinnamon Aroma Tea, Irises and Orchids Aroma Tea, Pervasive Aroma Tea, Jasmine Aroma Tea, Almond Aroma Tea, Sweet-scented Osmanthus Aroma Tea, Whitish Honey

余个品系，著名的有黄枝香、肉桂香、芝兰香、通天香、茉莉香、杏仁香、桂花香、白蜜香、蜜兰香等。不同名称的凤凰单丛茶，虽种在同一座乌岽山上，都制作成为条形、青蒂、绿腹、红镶边的茶叶，却因其树龄不一，种质有别，使每个单丛茶都具有自己的"山韵蜜味"，冲泡时，十步之遥便能闻到奇香，入口后，滋味酽醇，回味无穷。采收时，需按茶树不同植株所具有的不同风味，按类分别采摘和制造。由刚成熟的嫩梢制成的凤凰单丛茶，品质奇异。

Aroma Tea, and Honey Orchid Aroma Tea. Though all growing in the Wudong Hill and made into the strip-shaped tea with green stem, green central part, and red edge, the Dancong tea trees from Mt. Phoenix have different ages and different germ plasm and thus their own "Mountain Tastes and Honey Flavors". During brewing, a strong aroma can be smelt ten steps away. The tea soup has a strong taste that lasts long. Tea picking and making should be done differently according to the different flavors of different tea trees. The Dancong Tea from Mt. Phoenix made with the freshly mature tender tips has a unique quality.

- 凤凰蜜兰香单丛干茶、茶汤、叶底

Dry Tea, Tea Soup, Brewed Tea Leaves of Honey Orchid Aroma Dancong Tea from Mt. Phoenix

> 冻顶山乌龙茶

冻顶山乌龙茶主产于台湾省南投县鹿谷乡的冻顶山。相传清咸丰五年，南投鹿谷乡村民林凤池前往福建赶考，衣锦还乡时带回武夷乌龙茶苗36株种于冻顶山等地，之后逐渐发展成当今的冻顶茶园。冻顶山名为"冻顶"，却是四季常青，

- 冻顶山乌龙茶茶样、茶汤
  Sample and Soup of the Oolong Tea from Mt. Dongding

> Oolong Tea from Mt. Dongding

The oolong tea from Mt. Dongding is produced in Mt. Dongding of Lugu Township, Nantou County, Taiwan Province. According to the legend, Lin Fengchi, a villager from Lugu Township of Nantou County, went to Fujian Province to take the imperial examination in 1855 (the fifth year during the reign of Xianfeng in the Qing Dynasty). He succeeded in the examination and returned to his hometown an official. He brought back with him 36 seedlings of oolong tea trees from Mt. Wuyi, which were planted in Mt. Dongding and gradually developed into present-day Dongding tea garden. Though it is called Dongding (Freezing Summit), Mt. Dongding is green all the year round. It enjoys an annual average temperature of 20℃. Always covered with cloud and

年平均温度在20℃左右，常年云雾缭绕，日照相对短，红土质，天然环境可与武夷山媲美，最宜茶树生长。

冻顶茶一年四季均可采摘，但采集、制作极其讲究，鲜叶为青心乌龙等良种芽叶，其中以春茶最为名贵，被称为"黄金之叶"。

冻顶乌龙的发酵程度较轻，属半球形包种茶。其成茶外形呈半球形，紧结匀整，色泽墨绿油润，边缘有隐隐的金黄色。冲泡后的茶汤色泽金黄透亮，有浓郁的花香和熟果香，滋味浓醇干爽，入喉回甘，带有明显的焙火韵味，叶底边缘镶红边，中间淡绿，有青蛙皮般的灰白点。

mist, it receives relatively short sunshine. With red soil and a natural environment as good as that of Mt. Wuyi, it is the ideal place for growing tea trees.

The Dongding Tea can be picked all the year round. However, its picking and making should meet strict requirements. The fresh leaves are improved buds and leaves such as that of Qingxin (Green Heart) Oolong tea trees. The spring tea is the most precious one and is known as "Leaves of Gold".

Slightly fermented, the Oolong Tea from Mt. Dongding is a semisphere pouchong tea. The finished tea has a semisphere shape and is tight and even. It has a dark green and oily moist appearance, with slight golden color on the edge. The tea soup is golden, bright, and transparent, sending forth a strong flower scent and mature fruit scent. It tastes strong, mellow, and brisk, with a sweet aftertaste and an obvious baking flavor. The brewed tea leaves have red edge and light green middle part bearing grayish white spots like that on a frog's skin.

• 冻顶乌龙茶
Oolong Tea from Mt. Dongding

名水泡名茶
Famous Water Making Famous Tea

中国自古就有"好水泡好茶"的说法，水质能直接影响茶汤的品质。如果使用了水质不好的水，那么茶的清香甘醇就不能发挥出来。

中国人历来讲究泡茶用水，水要求活、甘、清、轻。中国人泡茶首选泉水，追求"好茶配佳泉"的完美组合。所以，在中国饮茶史上，特别是唐代以后有许多帝王将相、文人墨客、僧侣道人，为喝上一杯佳茗，不惜一切去远道汲取美泉。他们抒怀于胸，倾情于笔端，流传于百姓之口，为后人留下了许多有关茶水文化的遗迹。

There is an old saying in China—Good water makes good tea. Water quality can directly affect the quality of tea soup. Poor-quality water cannot give full play to the aroma and sweet mellowness of the tea.

Chinese people have always been attaching great importance to the water for brewing tea. They require the water to be live, sweet, clear, and light. In China, spring water is regarded as the best for brewing tea. Chinese people pursue the perfect combination of good tea and good spring water. In Chinese tea history, especially after the Tang Dynasty (618-907), many emperors, high-ranking officials, scholars, and monks spared no efforts to obtain satisfactory spring water from far-away places, simply for making a good cup of tea. They expressed their love for tea in poems and other literature works, which have been passed down from generation to generation. They also left behind many relics related to tea culture.

> 庐山谷帘泉

谷帘泉，又名"三叠泉"，位于江西省庐山最高处的汉阳峰西面，庐山三大峡谷之一的康王谷谷底，是庐山桃花源景区的主要景点。

• 谷帘泉
Gulian Spring

> Gulian (Grain Curtain) Spring in Mt. Lushan

Also known as Three-step Spring, the Gulian Spring is on the west side of the Hanyang Peak, the summit of Mt. Lushan in Jiangxi Province, and at the bottom of the Kangwang Valley, one of the three major valleys in Mt. Lushan. It is the main scenic spot of the Taohuayuan Scenic Resort in Mt. Lushan.

Mt. Lushan is formed with sand and stone and covered with dense vegetation. Water seeps through the vegetation, penetrates downwards through rock joint, passes the cracks in rock layer, and forms a green spring. It gushes out of the ravine and drops to the bottom of the valleys. The water quality is superb. The ancient people summed up eight merits of the Gulian Spring: pure, cool, fragrant, crystal-clear, gentle, sweet, clean, and smooth. However, it was Lu Yu, the Tea

庐山山体多由沙石组成，加之当地植被繁茂，而水通过植被，再沿岩石节理向下渗透，最后经过岩层裂缝，汇成一泓碧泉，从山涧喷涌而出，倾泻入谷。所以，水质特优。古人称谷帘泉有八大优点，即清、冷、香、冽、柔、甘、净、不噎人。但谷帘泉的出名，还得归功于唐代"茶圣"陆羽。据称陆羽当年为品评天下宜茶之水，长途跋涉，攀悬崖，过溪径，最终评定庐山康王谷水帘水为"天下第一名泉"。在泉水经流处的危崖之上，还建有一座单层四角四柱的观瀑亭，是观看谷帘泉的最佳处。为纪念陆羽评定康王谷水帘水，后人在亭下建了鸿渐（即陆羽）桥。在康王城前的山门口，建有一座四柱三门的牌坊，上书"天下第一泉"五个大字。入内，在山涧岩石之上，多有题刻，供人怀古。南宋理学家朱熹利用自己任当地地方长官的优势，在过观口山门前回马石旁的崖壁上，亲自用隶书写了"谷帘泉"三个大字。

Saint in the Tang Dynasty (618-907), who put the Gulian Spring on the map. After his painstaking tour to find the best water for brewing tea, Lu Yu concluded that the water from the Gulian Spring in Kangwang Valley of Mt. Lushan was the No.1 Spring in the World. Later, a one-floor Waterfall Viewing Pavilion with four corners and four pillars was built on the precipitous cliff near the spring. It was the best place to view the Gulian Spring. To commemorate Lu Yu's high evaluation of the Gulian Spring, local people built a Hongjian Bridge (also Lu Yu Bridge) under the pavilion. At the entrance to the mountainous Kangwang Town, an archway with four pillars and three gateways was built. On it were five Chinese characters meaning No.1 Spring in the World. In the town, there were many inscriptions on the mountainside rocks. Zhu Xi, an idealist philosopher in the Southern Song Dynasty (1127-1279), once served as the local magistrate. He personally wrote the name of the Gulian Spring in official script on the cliff wall in front of the mountain gate.

《煎茶水记》

中国历史上最早的关于煎茶论水的专著,作者是唐代的张又新。张又新,字孔昭,深州陆泽(今河北深州)人,唐代品茶家。唐元和九年进士第一名,历任江州刺史、左司郎中等。张氏擅长文辞,又善于品茶,尤其对煮茶用水颇有研究,约于公元825年,撰写《煎茶水记》。按书中内容严格说来,《煎茶水记》应为陆羽、刘伯刍和张又新共同创作。

Report on Water for Tea Cooking

This is the very first monograph on tea making in Chinese history. It was written by Zhang Youxin, also called Zhang Kongzhao, a native of Luze of Shenzhou (present-day Shenzhou City of Hebei Province) and a tea taster in the Tang Dynasty (618-907). Zhang became the first candidate in the highest imperial examination in the ninth year during the reign of Yuanhe in the Tang Dynasty (814). He later served as governor of Jiangzhou and a minister. Zhang was good at writing and tea tasting. He was an expert in tea making. In about 825, he wrote the *Report on Water for Tea Cooking*. However, the contents of the book also suggested that it could be a work jointly written by Lu Yu, Liu Bochu and Zhang Youxin.

- 茶
Tea

古代煮茶的方法

煮茶用水，山泉水最好，江水一般，井水最差。山泉水中，最好选取喷涌的泉水或穿越石池而慢流的水；奔涌急流的水不要饮用，长喝的话颈部会生病；江水要从离人活动区域远的地方汲取；井水要选人们经常汲水的井。

煮水时，冒出鱼眼大小的水泡，并微微作声，称作一沸；锅边有像泉涌一样的连珠泡，称作二沸；波涛翻腾的，称三沸。三沸是不能喝的老水。初沸腾时，可依据水的多少加些盐。二沸时，舀出一瓢水待用，用竹筷搅动沸水，再放入适量的茶末。待锅里茶水像惊涛翻涌并有水沫溅出时，立即缓缓倒入先舀出的那瓢水，以此止沸，并培育其中的"沫饽"，也就是水华。水华中薄的称"沫"，厚的称"饽"，细小轻盈的称"花"。分盛到每个碗里的汤沫要均匀。

煮好的茶须趁热喝完，放置时间过长的茶味道会非常暗淡。茶汤应为浅黄色，香气清馨，茶入口苦而咽下回甘。

Tea Cooking in the Past

Spring water is the best for cooking tea. The river water is the second best, followed by well water. Among spring waters, the one gushing out or slowly flowing through the rocky pond is the best. The water rushing out and flowing rapidly is of no good and will lead to neck ailment if drunk for a long time. Water from river should be taken at the places far away from human activities. If well water has to be used, the well from which people frequently take water should be chosen.

During cooking, when bubbles as big as the size of fish eye begin to emerge with small sizzling sound, it is the first boil. When bubbles appear in rapid succession from the rim of the kettle, it is the second boil. When the water rolls in the kettle, it is the third boiled. The third-boiled water is too old to be worth drinking. At the first boil, some salt can be added based on the quantity of the water. At the second boil, a ladleful of water should be taken and put aside for future use. Then, stir the boiling water with a bamboo stick and add proper amount of tea leaves. When the tea water begins to roll and some water foam spills out, immediately add in the ladleful of water taken moments ago to stop boiling and cultivate the Foam Dough, also known as Water Bloom. In fact, the thin substance in the water bloom is called Foam, the thick one is called Dough, and the fine and light one is called Bloom. The Foam should be evenly distributed to individual bowls.

The freshly cooked tea has the best taste. If put aside too long, the tea will taste lighter. The best tea soup should be light yellow with fresh aroma. The tea water tastes bitter at first but turns sweet after being swallowed.

> 济南趵突泉

趵突泉，位于山东省济南市城区的趵突泉公园内，在济南七十二泉中位居"第一泉"，因其泉水瀑流，跳跃如趵突而得名。趵突泉与大明湖、千佛山合称"济南三绝"。

泉水主要集中在一个东西长约30米，南北宽约20米的长方形泉池中。其内有三个大型泉眼，日夜涌水不息，势如鼎沸，状如堆雪。古人说这三注泉眼是东海之中的三座神山，即蓬莱山、方丈山和瀛洲山。趵突泉水质清净、甘洌，用来试水品茗，香正味醇。宋代文人曾巩选用趵突泉水品茗，盛赞它最宜煮茶。清代乾隆皇帝每到山东都选用趵突泉水品茗议政。

20世纪90年代以来，由于地下水的不断开发，趵突泉水逐年减

> Baotu (Spouting) Spring in Jinan City

The Baotu Spring is in the Baotu Spring Park of Jinan City, Shandong Province. Ranking first among the 72 springs in Jinan, it derives its name from its torrential spouting water. The Daming (Great Brightness) Lake, Mt. Qianfo (One-thousand-buddha Mountain), and Baotu Spring are called "Top Three Treasures of Jinan".

The spring is contained in a rectangular pond some 30 meters long from east to west and 20 meters wide from south to north. There are three spring mouths in the pond, from which, water gushes out day and night, making the pond look like a boiling cauldron or a plate holding the piling snow. The ancient people believed that the three spring mouths were the three divine mountains in the East China Sea, namely

少，使泉水"趵突"之观不能经常涌现。但用趵突泉水泡茶，特别是冲泡茉莉花茶，色如琥珀，清香扑鼻，实是佳茗配美泉之举。

Mt. Penglai, Mt. Fangzhang, and Mt. Yingzhou. The Baotu Spring has clear and sweet water, which, if used to cook tea, guarantees a right and mellow taste. In the Song Dynasty (960-1279), scholars liked to use the water from the Baotu Spring to cook tea and they hailed it as the best for tea making. In the Qing Dynasty (1616-1911), Emperor Qianlong always demanded the tea cooked with the water from the Baotu Spring every time he came to Shandong Province.

Since 1990s, with the nonstop exploitation of the underground water, the Baotu Spring has been producing less and less water year after year. The spouting water is no longer a frequent sight. However, the water from the spring remains the best for brewing tea, especially the Jasmine Tea. The tea soup has an amber color and sends forth a strong fresh aroma.

● 趵突泉 (图片提供：全景正片)
Baotu Spring

花茶

花茶，属于再加工茶，是以绿茶、红茶、乌龙茶等为原料，经过加窨各种香花而制成的茶叶品类。中国花茶的主要产地有福建、江苏、浙江、广西、四川、安徽、湖南、江西、湖北、云南等地。

花茶通常是根据窨花用的香花来定名，包括茉莉花茶、珠兰花茶、白兰花茶、桂花茶、玫瑰花茶等。每种花茶都具有香气鲜郁纯正、滋味浓醇鲜爽、汤色清亮艳丽的特点。

茉莉花茶是用茉莉鲜花窨制加工而成的再加工茶，是花茶中最广为人知的，其产量和品种也是众花茶中最多的一种。茉莉花茶的制作工序较为复杂，先将用来窨制的茶坯进行干燥和冷却处理，采摘来的新鲜茉莉花也要先摊晾及养护，然后用新鲜的玉兰打底调香，之后经过窨花拼合、通花散热、起花、烘焙、压花、提花等复杂工序制作而成。茉莉花茶外形条索紧细匀整，色泽黑褐油润，香气鲜灵持久，滋味醇厚鲜爽，汤色黄绿明亮，叶底嫩匀柔软，耐泡，高档茉莉花茶三泡之后仍有余香。

- 茉莉花茶
  Jasmine Tea

- 特级茉莉花茶干茶茶样、茶汤
  Sample and Soup of Superfine Jasmine Dry Tea

Scented Tea

The scented tea is a reprocessed tea made with the green tea, black tea, and oolong tea as the raw materials and by adding various scenting flowers. Chinese scented tea is mainly produced in Fujian, Jiangsu, Zhejiang, Guangxi, Sichuan, Anhui, Hunan, Jiangxi, Hubei, and Yunnan.

The scented teas are usually named after the scenting flowers added into it. There are Jasmine Tea, Zhulan Tree Tea, Gardenia Tea, Sweet-scented Osmanthus Tea, and Rose Tea. Each kind of scented tea has a fresh, strong, and pure aroma, a mellow and brisk taste, and a bright and beautiful tea soup.

The Jasmine Tea is a reprocessed tea added with fresh jasmine. It is the most famous among all scented teas. Its output and varieties are also the greatest among all scented teas. The Jasmine Tea is made through complicated processing procedures. First, the tea base for scenting is dried and cooled. The fresh jasmine is also air-cooled and conserved. Then, the fresh Magnolia flower is used for basic blending. The following procedures include scenting combination, cooling, flower lifting, baking, flower compressing, and flower extracting. The Jasmine Tea has tight, fine, and even strips and a blackish brown and oily moist appearance. Its aroma is fresh and long-lasting. It tastes mellow, fresh, and brisk. Its tea soup is yellowish green and bright. The brewed tea leaves are tender, even, and soft and last long after brewing. The top-grade Jasmine Tea keeps aroma after three rounds of brewing.

- 广西壮族自治区横县茉莉花种植基地
The Jasmine Planting Base in Hengxian County of Guangxi Zhuang Autonomous Region

> 崂山矿泉

崂山矿泉，位于山东省青岛市东北部的崂山风景名胜区内。崂山，人称"仙山"，据说秦始皇、汉武帝都曾到此炼丹求仙，以祈长生不老。宋代、元代以来，崂山是道人聚集之地，成为道教名山。这里道观林立，著名的道观有上清宫、下清宫、太平宫等，均是石壁瓦舍，深具道教玄机。因为崂山多奇峰削壁，形成了崂山33条清溪；而瑶池的九曲连环，还构成了幽深清邃的九水风光。其地到处是瀑布长鸣，山泉潺潺，山顶著名的金液泉、神水泉、玉液泉等名泉，与九水遥相辉映。

崂山处处水，汲汲皆可饮。这些矿泉，水质清澈，略带甘甜，又富含对人体有益的矿质元素，盛在

> **Mineral Spring in Mt. Laoshan**

Mt. Laoshan Scenic Resort in the northeast of Qingdao City, Shandong Province, produces mineral water. Also called the Fairyland Hill, Mt. Laoshan has been a Taoist stronghold since the Song and Yuan dynasties. In the earlier Qin and Han dynasties, the First Emperor of Qin and Emperor Wudi of Han had visited the mountain for making pills of immortality. Now, the mountain has countless Taoist monasteries. Among them the famous ones are the Palace of Celestial Freshness, Palace of Terrestrial Freshness, and Palace of Peace, all being typical Taoist buildings with rock walls and tiled roofs. The towering peaks and precipitous cliffs of Mt. Laoshan give rise to 33 clear streams. The winding water path of the Yaochi Pond forms the secluded nine-turn river landscape. Mt. Laoshan is full of

杯中，即使水面满出杯沿半米粒，也不致使水流外溢。因此，崂山矿泉有"神水""仙饮"之说。历史上，一些爱茶、写茶的文人学士，如唐代的李白，宋代的苏东坡，明代的文徵明，清代的顾炎

thunderous waterfalls and gurgling brooks. The famous springs on top of the mountain include the Gold Liquor Spring, Divine Water Spring, and Jade Liquor Spring. They chime in from afar with the nine-turn river.

Mt. Laoshan is full of water, clean and drinkable water. Its mineral springs are clear and sweet and contain rich mineral elements conducive to human health. Poured into a cup, the water from Mt. Laoshan won't overflow even if the water level is half a rice grain higher than the cup's edge. In fact, Mineral Spring in Mt. Laoshan is called the Divine Water and the Drink for the Deity. In history, some scholars who loved tea and wrote about tea had come to Mt. Laoshan for enjoying the tea made with its high-quality mineral water. These included Li Bai of the Tang Dynasty, Su Dongpo of the Song Dynasty, Wen Zhengming of the Ming Dynasty, Gu Yanwu and Wang Shishen of the Qing Dynasty, and Kang

• 山东崂山茶园（图片提供：全景正片）
A Mt. Laoshan Tea Garden in Shandong Province

- 好水烹好茶，美不可言
Good Water Brews Good Tea; the Beauty Is Beyond Words

武、王士禛，近代的康有为等，都曾慕名到崂山煮水试茗，开歌抒怀，留下众多吟茶墨宝。这里还是崂山春茶叶的产地，好水烹好茶，自然美不可言。

Youwei of modern times. Here, they enjoyed the tea and chanted poems in praise of the mountain and the tea. They left behind many well-known works about tea. Mt. Laoshan also produces Laoshan Spring Tea. Good water brews good tea; the beauty is beyond words.

> 无锡惠山泉

惠山泉，位于江苏省无锡市西郊惠山山麓锡惠公园内，泉因山而得名。惠山泉盛名始于中唐以后。品泉家刘伯刍根据其游历所至，分别排出宜茶水品七等，将无锡惠山寺泉水评定为"第二"。所以，惠山泉又称"二泉"。而陆羽则在更大范围内将天下宜茶水品评为二十等，也将惠山寺泉水评为"第二"。

> Huishan (Favorable Mountain) Spring in Wuxi City

The Huishan Spring is in the Xihui Park at the foot of Mt. Huishan (Favorable Mountain) in western suburb of Wuxi City, Jiangsu Province. The spring is named after the mountain. The Huishan Spring gained its fame since mid-Tang Dynasty. According to the spring taster Liu Bochu, the Huishan Spring in Wuxi City was the second best among the top seven springs good for brewing tea. Because of this, the Huishan Spring is also known as the Second Spring. What a coincidence, Lu Yu, in his larger-scope evaluation, also ranked the Huishan

• 天下第二泉：惠山泉
Huishan Spring, No.2 Spring in the World

惠山泉水为山水，是通过岩层裂缝过滤流淌的地下水，水质纯净，杂质极少，味甘质轻，是茶人心目中的泡茶美泉。唐代宰相李德裕曾用驿马装运惠山泉水进京烹茶。宋代时，从帝王将相到文人墨客都十分推崇惠山泉水，甚至不惜工本，将惠山泉水用舟车运载，送到京城开封，供达官贵人享用。清

Spring as the second best among the top 20 springs good for brewing tea.

The Huishan Spring originates in a mountain. It is filtered so fine by rock layers that it is pure, sweet, and light, containing little impurity. It is an ideal spring for brewing tea. Li Deyu, a prime minister in the Tang Dynasty, once had Huishan Spring water transported by post horses to the capital for brewing tea. In the Song Dynasty, the Huishan Spring was highly esteemed by emperors, princes, and scholars. They spared no efforts to transport Huishan Spring water to their capital Kaifeng, regardless of the cost. Emperor Qianlong of the Qing Dynasty conferred the title No.2 Spring in the World on the Huishan Spring. Zhao Mengfu, a master calligrapher in the Yuan Dynasty, wrote the Chinese

- 阿炳拉二胡雕像

阿炳，原名华彦钧，中国杰出的民间音乐家。《二泉映月》是他于20世纪30年代创作的一首二胡曲。这首曲子刻画出作者压抑在心底的幽愤、哀痛之情，同时还传达出面对生活艰辛绝不低头的倔强，极富艺术感染力。

A Statue of Abing Playing the Erhu Fiddle

Abing, or Hua Yanjun, was an outstanding folk musician of China. The erhu fiddle tune, *Moonlight on Second Spring*, was composed by him in 1930s. The tune expressed his resentment and sadness hidden in the depth of his heart. It also demonstrated his unyielding attitude towards the hardship in life. It is a tune full of artistic appeal.

朝乾隆皇帝御封其为"天下第二泉"。元代大书法家赵孟頫专为惠山泉书写了"天下第二泉"的匾额，至今仍完好地保存在泉亭后壁上。历代茶人都如此钟情惠山泉，为它创作诗歌加以咏叹。中国民间音乐家阿炳以惠山泉为素材创作的二胡曲《二泉映月》，至今仍广为流传。

characters of No.2 Spring in the World on a board for the spring. Today, the board still hangs on the rock wall behind the spring pavilion. All past tea lovers cherished the Huishan Spring and wrote poems and odes for it. The most famous work for the Huishan Spring is *Moonlight on Second Spring*, an erhu fiddle tune by a blind folk musician Abing. The tune is still popular nowadays.

- 惠山泉上池和中池
Upper Pond and Middle Pond of the Huishan Spring

- 惠山泉下池
Lower Pond of the Huishan Spring

惠山泉分上池、中池和下池：上池呈八角形，位于二泉亭内，水质最好；下池为不规则形，位于二泉亭前。惠山泉是地下水的天然露头，未受环境污染，再经过沙石过滤，汇集成流，水质自然清澈、晶莹。另外，由于水流通过山岩，泉水富含对人体有益的多种矿物质。

The upper pond, middle pond, and lower pond of the Huishan Spring: The octagon upper pond is in the Second Spring Pavilion and has the best water quality. The irregular lower pond is in front of the Second Spring Pavilion. The Huishan Spring is the natural outcrop of underground water and is free of pollution. Filtered by sand and stone, the spring water congregates into a crystal-clear stream. In addition, as the stream flows through rocks, it contains several minerals conducive to human health.

> 镇江中泠泉

中泠泉，位于江苏省镇江市金山以西的石弹山下，其名意为大江中心一股冰凉的泉水。古代时，由于中泠泉处于江心波涛汹涌的旋涡中，汲水颇不容易，只能选择在半夜和正午时，用由铜丸、壶身和壶

> Zhongling (Middle Cold) Spring in Zhenjiang City

The Zhongling Spring is under Mt. Shidan (Stone Pellet) to the west of Mt. Jinshan (Gold Mountain) of Zhenjiang City, Jiangsu Province. Its name means a stream of cold spring in the middle of a big river. In the past, people had great difficulty in taking water from the spring as it was in

● 镇江中泠泉 (图片提供：全景正片)
Zhongling Spring in Zhenjiang City

盖组成的水葫芦系在绳子上，沉入江心。当水葫芦正好深入到泉水窟中时，用绳子拉开壶盖，方可取得真正的中泠泉水。中泠泉属于由地下水沿石灰岩裂缝上涌而成的上升泉。泉水甘甜清冽，被唐代品泉家刘伯刍评为"天下第一"。

金山原是一个小岛，中泠泉也只是在枯水季时，当江水退潮后才有可能露出泉眼。后来随着长江主干道的北移，金山才与南岸相连。明清时，金山已成为旅游胜地，金山上有寺院，最著名的是金山寺，民间流传的白娘子"水漫金山寺"的故事就发生在这里。

the middle of a torrential river. They had to sink a roped water calabash composed of a copper ball, a kettle, and a lid into the center of the river at midnight or noon. When the water calabash reached the spring cave, they pulled the lid open to get a kettle of genuine Zhongling Spring water. The Zhongling Spring forms when the underground water rushes upwards through limestone cracks. Sweet and clear, it was assessed as the No.1 Spring in the World by Liu Bochu, the spring taster in the Tang Dynasty.

Mt. Jinshan was originally a small island. The Zhongling Spring showed its mouth only in dry season when river water ebbed. Later, with the main course of the Yangtze River moving northward, Mt. Jinshan began to connect with the southern bank. By the Ming and Qing dynasties, Mt. Jinshan had become a tourist resort. Among the many temples in Mt. Jinshan, the most famous is the Jinshan Temple. The folk legend, Flooding of the Jinshan Temple by Madam White Snake, happened here.

- 好茶需好水冲泡方能体现其真味
  A Good Tea Needs Good Water to Show Its True Taste

白蛇传的故事

　　白蛇传是中国民间家喻户晓的传说。故事讲述的是一条有千年修行的白蛇精，为了报答书生许仙前世的救命之恩，化作人形，取名白素贞并巧施妙计嫁与许仙，育有一子。在这个过程中受到同为蛇精的小青的帮助，但是婚后金山寺和尚法海告知许仙白素贞的真实身份，并施计将许仙骗至金山寺。白素贞与小青为救出许仙水漫金山寺，因触犯天条，白素贞被法海收入钵中，镇压于雷峰塔下。后来白素贞的儿子高中状元，将母亲救出，全家团聚。

The Story of Madam White Snake

The Story of Madam White Snake is a famous folk legend known to every Chinese household. Madam White Snake was a snake-turned spirit after one-thousand-year practice. Once, she was saved by a scholar Xu Xian. To repay him for his kindness, she turned into a human figure and named herself Bai Suzhen. With a clever plan, she married Xu Xian and gave birth to a son. In the process, she received help from Xiaoqing, another snake-turned spirit. However, Fahai, a monk from the Jinshan Temple told Xu Xian the real identity of his wife Bai Suzhen and lured him to the temple and retained him there. Bai Suzhen and Xiaoqing flooded the Jinshan Temple in an attempt to rescue Xu Xian. As her action violated the laws of God, Bai Suzhen was captured in a bowl by Fahai and held under the Leifeng Tower. Years later, Bai Suzhen's son became Number One Scholar after coming first in the highest imperial examination. He set free his mother and the family was reunited.

- 年画《水漫金山寺》
  A New Year picture: *Flooding the Jinshan Temple*

> 虎丘第三泉

虎丘第三泉，位于江苏省苏州市虎丘山下的虎丘寺中，为虎丘的胜景之一。虎丘，又名"海涌山"，因山形如蹲虎而得名。史籍上记载陆羽曾在虎丘寓居，发现虎丘泉水清冽甘甜，便在虎丘山上

> The Third Spring in Mt. Huqiu

The Third Spring is in the Huqiu Temple at the foot of Mt. Huqiu in Suzhou City, Jiangsu Province. It is one of the top scenic spots in Mt. Huqiu. Also called Mt. Haiyong, Mt. Huqiu derives its name from its shape like a crouching tiger (Hu: tiger; Qiu: hill). According to historical records, Lu Yu once lived in Mt. Huqiu and found that the Huqiu Spring was clear and sweet. He dug a well in Mt. Huqiu, which was later called Lu Yu Well. The Huqiu Spring was ranked the fifth best by Lu Yu and the third best by Liu Bochu. Therefore, the Huqiu Spring is also praised as the Third Spring in the World.

Now, the Third Spring in the Huqiu

- 虎丘第三泉
  The Third Spring in Mt. Huqiu

挖了一口井,世人称之为"陆羽井"。陆羽认为苏州虎丘寺泉水排到第五,而刘伯刍认为苏州虎丘寺泉水排到第三。于是,虎丘泉还有"天下第三泉"之美誉。

现存的虎丘寺泉,是一口古石井,井口方形,四面垒以石块。其虽存世1200年之久,但井水仍清澈可鉴,终年不枯。在第三泉附近,还有剑池、千人岩、冷香阁等古迹,虎丘塔、阖闾墓、二仙亭等景点,加之四周的山景,如同一幅美妙的山水画。

Temple is an ancient stone well with a rectangular mouth surrounded by rocks. Though over 1,200 years old, the well still produces clear water and remains watery all the year round. The well is surrounded by many scenic spots, including the ancient relics such as the Sword Pond, One-thousand-people Rock, and Pavilion of Cold Fragrance, and famous structures such as the Huqiu Tower, Tomb of King Helv, and Two-immortal Pavilion. Together with the surrounding mountainous views, the site is like a beautiful landscape painting.

- 《林榭煎茶图》【局部】文徵明(明)
The Painting of Tea Cooking in a Shed in Woods[Part], by Wen Zhengming (Ming Dynasty 1368-1644)

> 峨眉玉液泉

　　玉液泉，位于四川省峨眉山金顶之下的万定桥边、神水阁前，有"神水第一"的美誉。玉液泉四周，则是峨眉茶的产地。早在北宋时，诗人黄

- 峨眉山小景（图片提供：全景正片）
  A Glimpse of Mt. Emei

> Yuye (Jade Liquor) Spring in Mt. Emei

Hailed as the No.1 Divine Water, the Yuye Spring is beside the Wanding (All-Settled) Bridge and before the Divine Water Pavilion beneath the Golden Summit of Mt. Emei in Sichuan Province. The Emei Tea is produced in the area around the Yuye Spring. As far back as the Northern Song Dynasty (960-1127), poets Huang Tingjian and Su Dongpo visited the mountain and tasted the local tea brewed with local spring water. They left behind many literature works in praise of the local tea and spring water. Now, a stele before the Yuye Spring still bears the poems inscribed in different dynasties. The Emei Tea brewed with the water from the Yuye Spring has always been praised as the best tea-making recipe by scholars in the past dynasties.

　　Though over 1,000 years old, the

庭坚、苏东坡就曾来此咏泉品茗，留下了赞美玉液泉、峨眉茶的墨宝。如今在玉液泉前的一块石碑上，仍存有镌刻的历代诗文。玉液泉烹峨眉茶，是历代文人墨客所推崇的品茶之道。

　　玉液泉水虽历经千百年，仍大旱不竭，水品清澈明亮，光照鉴人。由于此泉不同凡响，所以，人们称它是"天上的神水""地下的甘泉"。这是一种极为难得的优质饮用矿泉水，除视觉、口感殊绝于众外，还含有微量的氡、二氧化硅等，对人体有很好的保健作用。

Yuye Spring still remains running even in severe drought. Its water is clear, bright, and refreshing. As an extraordinary spring, it has been praised as the Divine Water from the Heaven and the Underground Sweet Spring. It is indeed a rare high-quality mineral water for drinking. In addition to its outstanding appearance and taste, it contains small amount of radon and silicon dioxide and is conducive to human health.

• 好水冲泡出好茶
Good Water Brewing Good Tea

> 杭州虎跑泉

虎跑泉，位于浙江省杭州市西南大慈山白鹤峰下慧禅寺（俗称"虎跑寺"）侧院内。相传乾隆皇帝下江南时，喝过虎跑泉水泡的茶后，称其为"天下第三泉"。

- 虎跑泉
Hupao Spring

> Hupao (The Dreamed Tiger Pawed) Spring in Hangzhou City

The Hupao Spring is in the side courtyard of the Huichan (Wise Zen) Temple (also known as Hupao Temple) beneath the Baihe (White Crane) Peak of Mt. Daci (Great Mercy) in the southwest of Hangzhou City, Zhejiang Province. According to the legend, after tasting the tea brewed with the water from the Hupao Spring, Emperor Qianlong of the Qing Dynasty called the spring "the No.3 Spring in the World".

The water of the Hupao Spring gushes out after seeping through quartz sand. If poured into a bowl, the water won't overflow even if the water level is two to three millimeters higher than the edge of the bowl. The water is sweet and mellow and contains rich mineral substances that are conducive

虎跑泉水是从石英砂岩中渗出来的一股泉水，若将泉水盛于碗中，即便水面满出碗沿两三毫米，也不外溢。其水质甘洌醇厚，富含多种对人体有益的矿物质成分，是一种很珍贵的矿泉水。虎跑泉水与龙井茶叶合称"西湖双绝"，有"龙井茶叶虎跑水"之美誉。

虎跑泉附近还有建于唐代的虎跑寺、虎跑亭、滴翠轩等建筑，以及为纪念中国早期话剧活动家、艺术教育家李叔同在虎跑寺出家而建的弘一法师塔，它们与虎跑泉相映成趣，为品泉试茗增添了无限情趣。

to human health. It is indeed a precious mineral water. The Hupao Spring and the Longjing Tea are called "the Two Treasures of the West Lake". They are often mentioned together.

Close to the Hupao Spring are some buildings built in the Tang Dynasty (618-907), including the Hupao Temple, Hupao Pavilion, and Dicui (Dripping Green) Windowed Veranda. There is also the Master Hongyi Pagoda in honor of Li Shutong, one of the earliest stage play activists and art educators in modern China. He became a monk at the Hupao Temple. These buildings and the Hupao Spring contrast finely with each other and together add appeal to local tea culture.

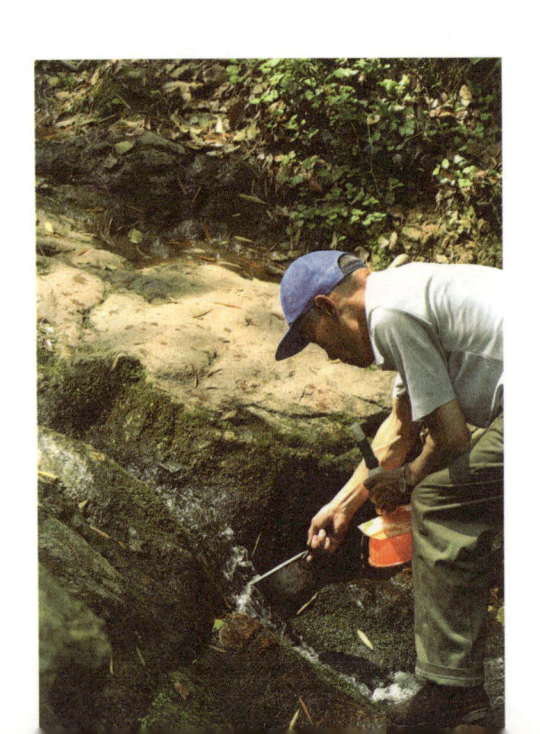

• 市民到虎跑公园打水
Local People Is Getting Water from the Hupao Park

> 杭州龙井泉

龙井泉，位于浙江省杭州市西湖西面风篁岭上，其西面就是盛产龙井茶的龙井村。龙井泉发现于三国时期，其闻名于世已有1700多年了。

龙井泉水出自山岩之中，由地下水与地面水两部分组成。当井中泉水溢出、井底泉水涌入时，水的比重和流速便不同，只要用小棍轻轻拨动水面，水面就会立刻出现一

- 老龙井
  Old Longjing Well

> Longjing (Dragon Well) Spring in Hangzhou City

The Longjing Spring is in Mt. Phoenix on the west of the West Lake, Hangzhou City, Zhejiang Province. The Longjing Village, the place of origin of the Longjing Tea, lies to its west. Discovered in the Three Kingdoms Period (220-280), the Longjing Spring has been well-known for over 1,700 years.

The water of the Longjing Spring comes from mountainous rocks. It consists of underground water and surface water. When water spills out of the well or when water gushes in from the bottom, an inward rippling division line will appear on water surface in response to the gentle stir by a small stick due to the differences in water specific gravity and flow speed. It is a wonder peculiar to this water. The Longjing Spring never dries out. Its water is sweet

条由外向内的波动分水线,是为奇观。龙井泉水的特点是四季不干,水味甘甜,清澈如镜。用龙井泉沏龙井茶,更是沁人肺腑。所以,历史上有"采取龙井茶,还烹龙井水"之说。

龙井泉的周围还有神运石、涤心沼、一片云等胜迹。在龙井泉的西侧,建有龙井寺茶室,用龙井泉水沏上在龙井村采制的龙井茶,别有一番情境。

此外,在离龙井泉几百米外的凤篁岭落晖坞,还有一口老龙井,它紧挨山岩,岩壁上有"老龙井"三字,也是一处与茶有关的重要人文景观。

and crystal-clear. The tea brewed with the water literally gladdens the drinker's heart and refreshes his mind. Therefore, an old saying goes that one should brew the Longjing Tea with none but the water from the Longjing Spring.

Around the Longjing Spring are several scenic spots such as the Divine Fate Stone, Clean Heart Marsh, and A Piece of Cloud. On the west side of the Longjing Spring is a Longjing Teahouse, which serves the Longjing Tea brewed with the water from the Longjing Spring.

In addition, there is an old Longjing Well in Luohui Dock beside Mt. Phoenix just several hundred meters from the Longjing Spring. The well is close to a mountain rock inscribed with three Chinese characters meaning Old Longjing Well. This is another important manmade relic related to tea.

- 龙井泉
  Longjing Spring

> 长兴金沙泉

金沙泉，位于浙江省长兴县顾渚山东麓。顾渚山是中国历史上最早的贡茶——顾渚山紫笋茶的产

> Jinsha (Gold Sand) Spring in Changxing City

The Jinsha Spring is located on the eastern foot of Mt. Guzhu in Changxing County of Zhejiang Province. Mt. Guzhu is the place of origin of the Zisun Tea, the earliest tribute tea in Chinese history. According to historical records, to get the true taste of the Zisun Tea from Mt. Guzhu, you have to cook the tea with the water from the Jinsha Spring, hence the saying of Guzhu Tea, Jinsha Water. After the Zisun Tea was chosen as a tribute tea, the water from the Jinsha Spring was also chosen as a tribute water. In the Tang Dynasty (618-907), when tributes were

- 顾渚山脚下的金沙泉
  The Jinsha Spring at the Foot of Mt. Guzhu

• 顾渚山皇家贡茶院遗址
The site of Royal Tribute Tea Station in Mt. Guzhu

地。史载，要尝到紫笋贡茶的真味，就非用金沙泉水煎茶不可。所以在历史上，有"顾渚茶，金沙水"之说。当紫笋茶被列为贡品后，又将金沙泉水列为贡水，在唐代，每年清明前进贡紫笋茶时，金沙泉水则用银瓶盛装，与紫笋茶一并送往京城长安（今陕西西安市）。

金沙泉三面环山，使泉水与外界地表水源隔绝。金沙泉眼正好处在花岗岩地层内，地表又为砾石冲积而成；加之，地面植被繁茂，竹林遍布，这种良好的自然环境和地质条件为金沙泉优质矿泉的形成创造了得天独厚的条件。

presented at the Qingming Festival, the water from the Jinsha Spring was bottled and sent to the capital Chang'an (present-day Xi'an City of Shaanxi Province) together with the Zisun Tea.

The Jinsha Spring is surrounded by mountains on three sides, which isolates it from surface water source. The mouth of the Jinsha Spring happens to be in the granite stratum and the land surface is made of alluvial gravels. In addition, the ground is covered with dense vegetation. Such ideal natural environment and geological condition guarantee the formation of the high-quality water of the Jinsha Spring.

名器配名茶
Famous Tea Set for Famous Tea

　　水是茶之母，器（茶具）是茶之父。茶具是茶文化的重要载体。茶具在中国古代时泛指制茶、饮茶时使用的各种工具，现在指与泡茶有关的专门器具。随着茶品种的不断增多和饮茶之风越来越盛行，茶具也经历了一个从无到有，从粗糙到精致的过程。

　　茶具的种类繁多，造型丰富，制作精良。根据材质的不同，茶具可分为陶质茶具、瓷器茶具、金属茶具、漆器茶具、竹木茶具、玻璃茶具等。

If water is the mother of tea, the utensil (tea set) should be the father of tea. Tea set is an important carrier of the tea culture. In ancient China, tea set referred to various utensils used in tea making and drinking. Now, it specifically refers to the utensils related to tea brewing. With the contents of tea drinking getting richer and richer, there have been more and more tea varieties and tea drinking has been more and more popular. Consequently, tea set has been developing from simple to sophisticated and from rough to delicate.

　　There are various kinds of tea sets. Exquisitely made, they are in all shapes and designs. Based on the materials, tea set can be divided into ceramic tea set, porcelain tea set, metallic tea set, lacquerware tea set, bamboo tea set, and glass tea set.

> 陶质茶具

陶质茶具是指用陶土烧制而成的茶具，是人类最早制作和使用的茶具之一，最初是粗糙的土陶，然后逐渐演变成比较坚实的硬陶，再后来发展为表面敷釉的釉陶，最后

> Ceramic Tea Set

Made of pottery clay, ceramic tea set was one of the earliest tea sets made and used by human. At first, the coarse pottery was used. Later, the more solid hard pottery took the place of it. Then, the glazed pottery and purple sand pottery prevailed. Ceramic tea set had played an important role in tea drinking history of human. However, it was gradually phased out by porcelain tea set. In recent years, ceramic tea set has regained some popularity with the rise of the tea art performance.

Of all ceramic tea sets, the purple sand tea set is the most famous. Made

- 紫砂壶
  Purple Sand Teapots

发展到紫砂陶。陶质茶具在人类饮茶史上发挥过重要的作用，但在瓷器茶具出现以后逐渐被淘汰。近年来，陶质茶具又随着茶艺表演的盛行而被重视起来。

陶质茶具中最著名的是紫砂茶具。紫砂陶以陶土为材料，含铁量高，有"泥中泥，岩中岩"之称。紫砂陶土质地细腻，颜色鲜艳，可塑性强。由于成陶火温需在1000℃~1200℃之间，所以成品致密，不渗漏，而表面光挺平整之中含有小颗粒状的变化，表现出一种砂质效果，又有肉眼看不见的气孔。这种结构的紫砂茶具有很好的吸附性和透气性，能吸附茶汁，蕴蓄茶味，且具有优良的宜茶性，即使是在炎热的夏天，壶中的茶叶依然可以隔夜不馊。

of pottery clay, the purple sand pottery contains rich iron. People praise it as the best of the clay and the best of the rock. It is refined, brightly colored, and highly ductile. As it should be made in a temperature between 1000℃ and 1200℃, the finished product is dense and leak-proof. Its surface is smooth and even, but with a sandy effect shown by some small particles. In fact, there are invisible pores on the purple sand pottery, which guarantee good adsorption and air permeability. It can adsorb tea soup and retain tea taste, making it a perfect tea set. Even in the hottest summer, the tea in the purple sand tea set can remain fresh after being left overnight.

- 紫砂如意云纹荷花茶具
  A Purple Sand Lotus-shaped Tea Set with Good-luck Cloud Design

> 瓷器茶具

瓷器茶具是用瓷土（主要是高岭土）烧制而成的茶具，是在陶质茶具之后发明和使用的茶具。瓷器的烧制温度比陶土要高，在1200℃左右，胎体坚固致密；瓷器表面施釉，无吸水性，容易清洗，没有异味；瓷器的造型美观、装饰精巧，可以很好地保留茶的色、香、味；瓷器的保温度适中，不烫手，不炸裂，是茶具中使用最广泛的品种。

根据施釉色彩的不同，瓷器茶具可以分为青瓷茶具、白瓷茶具、黑瓷茶具和彩瓷茶具四种。

> Porcelain Tea Set

Porcelain tea set is made of porcelain clay (mainly kaolin). It was invented and used after ceramic tea set. Porcelain requires a higher baking temperature than ceramics. At 1200℃, the porcelain body becomes solid and compact. Coated with glaze, the porcelain surface does not absorb water or give any peculiar smell and can be easily washed clean. Porcelain has beautiful design and exquisite decoration and can better retain the color, aroma, and taste of the tea. In addition, porcelain can properly preserve heat. It does not burn your hand or crack under heat. It is the most widely used tea set.

Based on the color of the glaze, porcelain tea set can be divided into celadon tea set, white porcelain tea set, black porcelain tea set, and faience tea set.

Celadon Tea Set

Coated with green glaze, celadon is a major type of porcelain produced in China. Celadon was first produced in the Eastern Han Dynasty (25-220) and the product at that time had pure color and a transparent and luminous appearance. In the Jin Dynasty (265-420), Yue Kiln, Wu Kiln, and Ou Kiln had been in large scale. In the Song Dynasty (960-1279), the celadon tea set produced in Ge Kiln in Longquan, Zhejiang Province, one of the top five kilns at that time, reached its zenith and was sold to many places far and near. In the Ming Dynasty (1368-1644), celadon tea set gained great fame at home and abroad for its fine and smooth texture, stately design, glistening glaze, and elegant pattern.

White Porcelain Tea Set

White porcelain grows out of celadon and is one of the traditional Chinese porcelains. In the Tang Dynasty (618-907), white porcelain was called Fake White Jade. Since the Tang Dynasty (618-907), many kilns have been engaged in producing white porcelain tea set. Among them are Xing Kiln in Renqiu of Hebei Province, Yue Kiln in Yuyao of Zhejiang

- 青瓷茶具
A Celadon Tea Set

青瓷茶具

青瓷施青釉，是中国瓷器生产的主要品类。东汉时期已开始生产色泽纯正、透明发光的青瓷。晋代时越窑、婺窑、瓯窑已具相当规

- 青瓷盏托（宋）
A Celadon Saucer (Song Dynasty 960-1279)

模。宋代，作为当时五大名窑之一的浙江龙泉哥窑生产的青瓷茶具，已然达到鼎盛时期，远销各地。明代，青瓷茶具更以质地细腻、造型端庄、釉色青莹、纹样雅丽而蜚声中外。

白瓷茶具

白瓷是由青瓷发展而来的，是中国传统瓷器的一种，在唐代时白瓷有"假白玉"之称。自唐以来生产白瓷茶具的窑场很多，如河北任丘的邢窑、浙江余姚的越窑、湖南的长沙窑、四川的大邑窑等，但最为著名的当属江西景德镇出产的白瓷茶具，以"白如雪、薄如纸、明如镜、声如磬"而闻名于世。白瓷茶具色泽晶莹洁白，造型精巧典雅，极具艺术欣赏价值。同时，其纯白质地的茶具更能衬托出各种茶

- 白瓷茶碗（唐）
A White Porcelain Bowl (Tang Dynasty 618-907)

Province, Changsha Kiln in Hunan Province, and Dayi Kiln in Sichuan Province. The most famous is the white porcelain tea set produced in Jingdezhen of Jiangxi Province, which has been well-known worldwide. white porcelain from Jingdezhen is as white as a snow, as thin as paper, and as clear as a mirror. In addition, its sound is as pleasing to the ear as the chime stone. With a sparkling white color and delicate and graceful design, white porcelain tea set has a high value in artistic appreciation. Meanwhile, its pure white texture can better set off the color of various tea soups. Finally, it performs well in heat transmission and preservation. It is indeed a rare utensil for tea drinking.

Black Porcelain Tea Set

Black porcelain is coated with black glaze. It appeared in the Late Tang Dynasty and gained great fame in the Song Dynasty due to the tea contest popular at that time. During the contest, the judges first examined the color and evenness of the tea soup bloom on the surface. The fresh white one was the best. Then, they examined whether there was water trace between the soup bloom and the teacup wall and when it appeared.

• 白瓷茶具
A White Porcelain Tea Set

汤的色泽，可谓相得益彰，加之传热、保温性能适中，故堪称饮茶器皿中的珍品。

黑瓷茶具

黑瓷施黑釉，始于唐代晚期，因宋代时兴斗茶而驰名。宋人斗茶，一看茶面汤花色泽和均匀度，以"鲜白"为佳；二看汤花与茶盏相接处水痕的有无和出现的迟早，以"盏水无痕"为上。黑瓷茶盏最适宜斗茶，因此在宋代时是瓷器茶具中最多的品种。福建建窑、江西吉州窑、山西榆次窑等都是黑瓷茶具的主要产地，其中以建窑生产的建盏最为人称道。

The tea producing no water trace was the best. Black porcelain teacup was the best for tea contest. Therefore, it was the dominant porcelain tea set variety in the Song Dynasty (960-1279). Black porcelain tea set was mainly produced by Jian Kiln in Fujian Province, Jizhou Kiln in Jiangxi Province, and Yuci Kiln in Shanxi Province. Among them, Jian teacup produced by Jian Kiln was praised as the best.

Faience Tea Set

There are many kinds of faience tea sets. Among them, the blue and white porcelain tea set, especially that produced in Jingdezhen City of Jiangxi Province, is the best and the most famous. In

• 黑釉木叶纹盏（宋）
A Black Porcelain Teacup with a Leaf Pattern (Song Dynasty 960-1279)

ancient China, black, blue, and green were all referred to as blue. Therefore, the meaning of blue and white was more extensive at that time. The great feature of blue and white porcelain is its blue and white grains. With the porcelain-making techniques getting better and better, new varieties of faience have appeared, including the Red in Blue and White Glaze and the Blue and White with Overglaze Colors.

- 黑瓷茶具
  A Black Porcelain Tea Set

彩瓷茶具

彩瓷茶具的花色品种很多，以青花瓷最为著名，其中江西景德镇的青花瓷首屈一指。中国古代将黑、蓝、青、绿等颜色统称为"青"，所以"青花"的概念比现在广泛。青花瓷是以花纹蓝白为最大的特点，随着制瓷技术的提高，彩瓷中出现了青花釉里红、斗彩等器具。

- 彩瓷茶具
  A Faience Tea Set

> 金属茶具

　　金属茶具是指由金、银、铜、铁、锡等金属材料制作而成的茶具。在先秦时期，青铜器就得到了广泛应用。南北朝时期，出现了金银质的饮茶器皿，至隋唐时期金属茶具制作达到高峰。但是随着陶瓷茶具的兴起，金属茶具从明代

> Metallic Tea Set

Metallic tea set is made of metallic materials such as gold, silver, copper, iron, and tin. In the pre-Qin period, bronze ware was widely used. In the Northern and Southern dynasties, gold and silver tea set appeared. In the Sui and Tang dynasties, making of metallic tea set reached its peak. However, with the rise of ceramic tea set, metallic tea set gradually disappeared since the Ming

- 摩羯纹蕾纽三足盐台（唐）
  古人煎茶时放盐、胡椒等佐料的茶具。
  A Tripod Salt Container with Male Goat Pattern and Bud-shaped Knob (Tang Dynasty 618-907)
  This is a utensil for containing condiments such as salt and pepper used by the ancient people during tea cooking.

开始就逐渐消失。人们虽然认为金属茶具，尤其是用锡、铁等制作成的茶具煮水泡茶会使茶味走样，但是又看中它密闭性好，具有防潮、避光的性能，因而也常用来贮藏散茶。

Dynasty. Although people think that metallic tea set, especially those made of tin and iron, will make the tea soup taste strange, it is still used by people for storing bulk tea because of its satisfactory sealing ability and good performance in preventing dampness and avoiding light.

搪瓷茶具

搪瓷茶具是在金属表面附以珐琅层的茶具，多以钢铁、铝等为坯胎，涂上一层或多层珐琅浆，经干燥、烘烤烧制而成，坚固耐用，轻便耐腐蚀。搪瓷起源于埃及，于元代传入中国。明代时，中国创制了镶嵌工艺品景泰蓝茶具，制作精美，色彩艳丽，具有鲜明的民族特色。20世纪初，中国开始大量生产搪瓷茶具。

搪瓷茶具种类多样，有仿瓷茶杯、保温茶杯、网眼茶杯、搪瓷茶盘等。搪瓷茶具质轻、易洗涤、耐高温、耐酸碱腐蚀，但由于传热较快，容易烫手，也容易烫坏桌面。

Enamel Tea Set

Enamel tea set has an enamel layer on its metallic surface. Normally, one or several layers of enamel plasm are applied to tea set's steel, iron, or aluminum body. Then, the tea set is dried and baked into finished product, which is solid, durable, light, and corrosion-resistant. Enamel originated from Egypt and was spread to China in the Yuan Dynasty (1206-1368). In the Ming Dynasty (1368-1644), cloisonne tea set was invented in China. This mosaic artwork is exquisitely made and brightly colored and has distinct characteristics of the Chinese nation. By the beginning of the 20th century, China began producing enamel tea set in large quantities.

There are diversified varieties of enamel tea set, including porcelain-like teacup, thermos teacup, teacup with a filter, and enamel tea tray. Enamel tea set is light and easy to wash and performs good in resisting high temperature and acidic and alkali erosion. However, enamel transmits heat quickly and may scald hand and table top.

• 搪瓷茶具
An Enamel Teapot

> 漆器茶具

漆器茶具是采用天然漆树汁液进行炼制，掺进所需色料而制成的茶具。中国的漆器起源甚早，在7000年前的河姆渡文化遗址中就发现了木胎漆碗，但到了清代才以脱胎漆器作为茶具使用。漆器茶具

> Lacquer Tea Set

Lacquer tea set is made with natural lacquer tree sap added with the desired pigment. Lacquerware had a long history in China. Wood-based lacquer bowls were found from the Hemudu Culture Relics dating back to 7,000 years ago. However, bodiless lacquerware was not used as tea set until the Qing Dynasty (1616-1911). Lacquer tea set is black, moist, delicate, bright-colored and dazzling, and clear as a bright mirror. Involving the art of calligraphy and painting, lacquer tea set has both practical

- 漆器茶具
A Lacquer Tea Set

脱胎漆茶具

脱胎漆茶具多为黑色，也有黄棕、棕红、深绿等颜色，轻巧美观，色泽光亮，耐水浸、耐高温、耐酸碱腐蚀等。其制作十分复杂，先按照茶具的设计要求做出木胎或泥胎模型，用布或绸料漆裱，再上几道漆灰料，然后脱去模型，再经填灰、上漆、打磨、装饰等多道工序，才能制成典雅的脱胎漆茶具。

Bodiless Lacquer Tea Set

Most bodiless lacquer tea sets are black. There are also the ones in yellowish brown, brownish red, and dark green. They are light, beautiful, and bright, and performs well in enduring water immersion and resisting high temperature and acidic and alkali erosion. Bodiless lacquer tea set is made through a complicated procedure. First, a wooden or earthen model set is made based on the design requirements of the tea set. Then, it is painted with cloth or silk fabric and applied with several layers of paint ash. Next, the model set is taken off, which is followed by several other steps such as ash filling, painting, grinding, and decoration. After all these steps, an elegant bodiless lacquer tea set is made.

乌润轻巧，光彩夺目，明镜照人，又融书画艺术于一体，既有实用价值，还有艺术欣赏价值。较为著名的漆器茶具有北京雕漆茶具，福州脱胎茶具，江西鄱阳、宜春等地生产的脱胎茶具。

value and artistic appreciation value. Among the famous lacquer tea sets are Beijing carved lacquer tea set and the bodiless lacquer tea set made in Fuzhou City of Fujian Province, and Poyang County and Yichun City of Jiangxi Province.

> 竹木茶具

　　竹木茶具是指用天然竹木制成的茶具。竹木茶具历史悠久，早在隋唐以前就已出现，当时除陶瓷器外，民间茶具多用竹木制作而成。陆羽在《茶经·四之器》中列出的28种茶具，多数是用竹木制作的。清代时，在四川出现了一种竹编茶具，由内胎和外套组成，内胎多为陶瓷类饮茶器具，外套用精选慈竹，经劈、启、揉、匀等多道工序，制成粗细如发的柔软竹丝，经烤色、染色，再按茶具内胎形状、

> Bamboo-Wood Tea Set

Bamboo-Wood tea set is made of natural bamboo. It had a long history in China and first appeared before the Sui Dynasty (581-618). At that time, besides the ceramic and porcelain utensils, most folk tea sets were made of bamboo-wood. In *Chapter Four: Tea Set of his Tea Sutra*, Lu Yu listed out 28 kinds of tea sets and most of them were made of bamboo. In the Qing Dynasty (1616-1911), a bamboo-woven tea set appeared in Sichuan Province, which was composed of an inner body and an outer casing. The inner body was normally a ceramic or porcelain tea drinking utensil. The outer casing was made of sinocalamus affinis (a kind of bamboo). Through chopping, opening, rubbing, and evening, the soft bamboo threads thick as human hair were made. After baking and dying, the threads were woven into a casing for the inner body based on its shape and size. Together, they formed an integrated tea

- 竹木茶具
  A bamboo-wood tea set

大小编织嵌合，使之成为整体如一的茶具。这种茶具泡茶后不易烫手，外套能有效地保护内胎。竹编茶具既是一种工艺品，又富有实用价值，主要品种有茶杯、茶盅、茶托、茶壶、茶盘等，多为成套制作。

set. Such tea set won't scald hand and its outer casing can effectively protect the inner body. Bamboo-woven tea set is an art work that has practical values. Its major varieties include teacup with and without handle, saucer, teapot, and tea tray. They are often made as a complete set.

良茶佳具

适宜冲泡绿茶的茶具

绿茶的本质特征是"水清茶绿"，对茶具要求讲求变化。

冲泡绿茶在色、香、形、味上都有讲究，好的绿茶，宜选用玻璃杯或白瓷盖碗。玻璃杯宜赏形，可以清楚地观察到绿茶在水中缓缓舒展、游动、变化的过程，适合针形茶、扁形茶；白瓷茶具宜赏茶汤，用细腻的白瓷来衬托绿茶茶汤嫩绿明亮

- 碧螺春适宜用白瓷茶具冲泡
  White Porcelain Tea Set Is Suitable for Brewing Biluochun Tea

的颜色，非常赏心悦目，适合碧螺春、信阳毛尖等细嫩显毫的揉捻茶。

适宜冲泡青茶的茶具

青茶，分为闽南乌龙、闽北乌龙、广东乌龙、台湾乌龙四大支系。各支系冲泡手法同中有异，对茶具的要求也是有所区别的。

闽北乌龙，以独特的"岩韵"著称，代表品种有大红袍、白鸡冠、水金龟、铁罗汉、水仙、肉桂等。适合用宜兴的紫砂壶冲泡，因紫砂能吸取茶中部分火气。另外，白瓷盖碗也是较为常见的冲泡岩茶的器具。

闽南乌龙，代表品种有铁观音、黄金桂等，适合用白瓷器皿冲泡，紫砂茶具也与之相称。

广东乌龙，主要代表是凤凰单丛，适合用紫砂壶冲泡，也可以用白瓷茶具赏形、观汤色。

台湾乌龙，代表品种有冻顶茶、阿里山茶等，属于发酵较轻的茶类，宜用台湾本地所产的瓷器冲泡。

适宜冲泡白茶的茶具

冲泡白茶的茶具宜用古朴自然的陶器、石器、木器茶具，忌豪华奢侈的器皿。此外，白毫银针可用玻璃器皿赏形，以无色无花的直筒形透明玻璃杯为好，这样可从各个角度欣赏到杯中茶的形和色，以及它们变幻的姿态。但是，用玻璃器皿冲泡有损茶味。

适宜冲泡黄茶的茶具

冲泡黄茶所用器具以紫砂为佳。此

• 凤凰单丛适宜用紫砂壶冲泡
The Purple Sand Teapot is Suitable for Brewing Phoenix Dancong Tea

外，黄茶也适合用玻璃器皿赏形。如君山银针在冲泡时，茶姿的形态、茶芽的沉浮、气泡的生成等，都是其他茶品冲泡时罕见的。茶芽在透明的玻璃杯中上下浮动，个个竖立，有"三起三落"之风采。

适宜冲泡红茶的茶具

红茶代表品种有祁门红茶、滇红金毫、正山小种等，因茶形各异，所以对茶具要求也有不同。

祁红具有"宝光、金晕、汤色红艳"三大特点，适合用白瓷冲泡，用玻璃器皿赏汤。祁红还是备受英国皇室青睐的茶饮，所以用一套斗彩绘花、吞金掐银的茶具，调入鲜奶、柠檬，再配上三明治与松饼，便是一道纯正的英式下午茶。

滇红适合用玻璃器皿赏茶形，也可以选用三才杯。

正山小种适宜选用紫砂茶具或陶质茶具冲泡，可以抵消部分松烟气，使香气更加纯正，滋味更加醇厚。

适宜冲泡黑茶的茶具

黑茶的代表茶品有云南的普洱茶、广西的六堡茶、湖南的春尖等。黑茶可以冲泡，也可以煎煮，适合选用陶制茶具或砂粒较粗的紫砂茶具，借砂土的吸附性除去茶叶在存放过程中形成的不好味道，突出黑茶的品质。

Good Tea with Good Tea Set

Tea Set Suitable for Brewing the Green Tea

Green tea features clear water and green tea leaves. The tea set for brewing green tea should be changed according to different varieties.

- 祁红适宜用白瓷茶具冲泡
White Porcelain Tea Set is Suitable for Brewing the Keemun Black Tea.

Brewing of green tea should meet the requirements on color, aroma, appearance, and taste. For a high-quality green tea, a glass or a white porcelain tureen is a good brewing utensil. Use of glass is for enjoying the shape change of the green tea leaves, including its slow unfolding, sliding, and changing in the water. Glass is the best choice for needle-shaped tea and flat tea. Use of white porcelain tea set is for enjoying the color of the tea soup. The fine and smooth white porcelain can better set off the tender and bright greenness of the tea leaves and make the drinking process also pleasing to the eye and the mind. White porcelain tea set is the best choice for the rolled tender tippy teas such as Biluochun Tea and Maojian Tea from Xinyang.

Tea Set Suitable for Brewing the Oolong Tea

There are four major oolong tea branches, namely South Fujian Oolong, North Fujian Oolong Tea, Guangdong Oolong Tea, and Taiwan Oolong Tea. Each branch has its own brewing techniques and unique requirements on tea set.

North Fujian Oolong Tea is known for its unique charm of rock. The representatives of this branch include Dahongpao (Big Red Robe) Tea, Baijiguan (White Comb) Tea, Shuijingui (Water Golden Turtle) Tea, Tieluohan (Iron Arhat) Tea, Shuixian (Narcissus) Tea, and Rougui (Cinnamon) Tea. They'd better be brewed with purple sand teapot from Yixing City, because purple sand can take some heat away from the tea. Besides, white porcelain tureen is also a commonly-used tea set for brewing rock tea.

The representatives of South Fujian Oolong are Tieguanyin Tea and Huangjingui (Golden Osmanthus) Tea. They'd better be brewed with white porcelain tea set or purple sand tea set.

Guangdong Oolong Tea is represented by Phoenix Dancong Tea. It had better be brewed with the purple sand teapot. White porcelain tea set can also be used to enjoy the shape of the tea leaves and the color of the tea soup.

The representatives of Taiwan Oolong Tea include Dongding Tea and Mt. Ali Tea, both being slightly-fermented teas. They'd better be brewed with the porcelain tea set produced in Taiwan.

• 普洱茶适宜用陶质茶具冲泡
Ceramic Tea Set is Suitable for Brewing the Pu'er Tea.

Tea Set Suitable for Brewing the White Tea

The natural and unsophisticated ceramic tea set, stone tea set, and wooden tea set are suitable for brewing the white tea. The luxurious utensils should be avoided. In addition, the White-tipped Yinzhen (Silver Tip) Tea can be brewed with glass utensils for shape appreciation. The best is the tube-shaped transparent glass without any pattern on it. Such glass guarantees a clear view of the tea's shape, color, and endless change from different angles. However, glass tea set will compromise tea taste to certain extent.

Tea Set Suitable for Brewing the Yellow Tea

The purple sand tea set is the best for brewing the yellow tea. In addition, glass utensil is also a good choice for appreciating the shape of the tea leaves. For example, when the Yinzhen (Silver Tip) Tea from Mt. Junshan is brewed, the changing shape of the tea leaves, the rising and sinking of the tea buds, and the formation of the bubbles are all worth viewing. They are rare with other teas. The tea buds move up and down in the transparent glass and each stands vertically like a tree. It is a mesmerizing scene in its own right.

Tea Set Suitable for Brewing the Black Tea

The representatives of the black tea include Keemun Black Tea, Dianhong Jinzhen(Golden Tip) Tea, and Zhengshan Souchong Tea. The different tea shapes result in different requirements on tea set.

　　The Keemun Black Tea features treasure light, gold halo, and brilliant red tea soup. It had better be brewed with a white porcelain tea set and its soup appreciated with glass utensils. The Keemun Black Tea is loved by British royal family. Brewing it in a tea set with overglaze flower pattern and gold and silver decoration, adding fresh milk and lemon juice into it, and preparing sandwiches and muffin, you have a genuine English-style afternoon tea.

　　Glass utensils are suitable for appreciating the shapes of Dianhong (Yunnan) Black Tea. Sancai Cup is also a good choice.

　　The purple sand tea set or ceramic tea set is suitable for brewing Zhengshan Souchong. Such tea set can offset part of the pine smoke flavor and make tea aroma purer and tea taste mellower.

Tea Set Suitable for Brewing the Dark Tea

Representatives of the dark tea include the Pu'er Tea of Yunnan Province, Liubao (Six-Cattle) Tea of Guangxi, and Chunjian (Spring Tip) Tea of Hunan Province. The dark tea can be brewed or cooked. The ceramic tea set or purple sand tea set with coarser sand grains is suitable for brewing the dark tea. The adsorptive sand can help remove the undesired taste formed during tea storage and accentuate the features of the dark tea.

如何泡好茶
How to Make a Good Cup of Tea

　　茶的真香本味、品质高低，必须通过正确的冲泡和品尝才能展现出来。中国茶种类多样，每一种茶都有自己独特的风味，因此不同的茶类有不同的冲泡方法，只有了解茶的特性，掌握了合理的冲泡程序，并经过反复实践，才能冲泡出美味的中国茶。泡茶不仅可以满足人们的物质需求，更能让人在泡茶的过程中修身养性、陶冶情操。

The true aroma, taste, and quality of the tea can be displayed only through correct brewing and tasting. There are diversified Chinese teas and each has its unique flavor that requires a unique way of brewing. Only by understanding the characteristics of the teas, grasping the reasonable brewing procedures, and practicing them repeatedly can one succeed in making a good cup of Chinese tea. Tea brewing can not only meet people's material demands, but more importantly, cultivate their moral characters and mould their temper in the process.

> 冲泡绿茶

冲泡绿茶一般选用容水量在100～150毫升的玻璃杯，茶叶的投放量为2～5克，冲入开水150毫升左

> Brewing Green Tea

The green tea is normally brewed with a 100ml to 150ml glass. A glass of tea needs two to five grams of tea leaves and 150ml boiling water. For the fine and tender high-quality tea, more tea leaves can be used depending on the drinker's need.

The green tea can be brewed in various ways. The commonest ways include Top Throw, Middle Throw and Bottom Throw. Different green teas have different characteristics and require different brewing methods. The upper

• 绿茶的上投法

先用热水温杯，提高茶具的温度，再向杯中注热水；等水温适度时，再向杯中投放茶叶。这样冲泡出来的绿茶保存了极好的嫩度，叶片依次展开，徐徐下沉，非常美妙。

Top Throw Method of the Green Tea

First, warm up the teacup with hot water. Then, pour in hot water, wait for the right temperature, and put in tea leaves. In this way, the tenderness of the green tea is best preserved. The tea leaves will unfold and sink slowly one by one, a beautiful scene to enjoy.

- **绿茶的中投法**

先用热水温杯，然后向杯中注入1/3杯70℃～85℃的热水；再将适量茶叶投入杯中，让茶叶吸足水分舒展开来；最后以悬壶高冲法向杯中注热水，让茶中的可溶性物质尽快浸出。这样冲泡出来的茶叶才会适时展开，嫩绿的茶芽呈现出来，口感清新浓郁。

Middle Throw Method of the Green Tea

First, warm up the teacup with hot water. Then, pour in hot water between 70°C and 85°C to fill one third of the cup, put in proper amount of tea leaves, and let them absorb water and fully unfold. Finally, pour in hot water from a highly lifted teapot to free the soluble substances from the tea as quickly as possible. In this way, the tea leaves can timely unfold to show the tender tea buds, guaranteeing a fresh and strong tea taste.

右。细嫩的名优茶用量可视品饮者的需要稍微多一点。

绿茶的冲泡方法多种多样，常见的有上投法、中投法和下投法。由于绿茶的品性不同，所以冲泡方法有所不同。上投法多适用于碧螺春、信阳毛尖、蒙顶甘露等细嫩度极好的绿茶；中投法适用于黄山毛峰、庐山云雾等细嫩、紧实的名优绿茶；下投法适用于西湖龙井、竹

application applies to the green tea with superb tenderness such as Biluochun, Maojian, and Mengding Ganlu. The medium application applies to the tender and compact high-quality green tea such as the Maofeng Tea from Mt. Huangshan and Yunwu Tea from Mt. Lushan. The lower application applies to the high-quality green tea with fat and strong buds and leaves such as the West Lake Longjing Tea, Zhuyeqing Tea, Guapian

叶青、六安瓜片、太平猴魁等芽叶肥壮的名优绿茶。另外也可根据个人喜好来选择适宜的冲泡方法。

Tea from Lu'an City, and Houkui Tea from Taiping. In addition, brewing method can also be chosen based on the drinker's personal likes and dislikes.

● **绿茶的下投法**

先用热水温杯后，将适量茶叶投入杯中；然后向杯中注入70℃~85℃的热水；将茶杯沿逆时针方向转动数圈，让茶叶和水充分接触。这种冲泡方法可激发出绿茶的色、香、味。

Bottom Throw Method of the Green Tea

First, warm up the teacup with hot water. Then, put proper amount of tea leaves into the cup and pour in hot water between 70℃ and 85℃. Turn the teacup anticlockwise for several circles to let the tea leaves fully contact with the water. This brewing method can fully release the color, aroma, and taste of the green tea.

绿茶茶艺——西湖龙井茶艺

西湖龙井是绿茶中最具特色的茶品之一，被誉为"绿色皇后"。

Tea Art for the Green Tea—Tea Art for the West Lake Longjing Tea

The West Lake Longjing is one of the most characteristic green teas. It has been praised as the Queen of the Green teas.

- 备具

冲泡龙井茶要高档无花的玻璃杯，以便更好地欣赏茶叶在杯中翩翩起舞的情景，观赏碧绿的汤色。

Tea Set Preparation

The Longjing Tea should be brewed with a top-grade glass without pattern on it, which allows free appreciation of the dancing tea leaves and the green tea soup.

- 赏茶

龙井茶外形扁平光滑，享有色绿、香郁、味醇、形美"四绝"之盛誉。优质龙井茶以清明前采制的为最好，称为"明前茶"。

Tea Appreciation

The Longjing Tea has a flat and smooth appearance. It is known for its four beauties: greenness, rich aroma, mellow taste, and beautiful appearance. The high-quality Longjing Tea picked before Pure Brightness, known as pre-PB tea, is the best.

- 鉴水

冲泡西湖龙井以杭州虎跑泉水为最佳，可起到相得益彰的效果。可请宾客品赏这甘甜清冽的佳泉。

Water Appreciation

The water from the Hupao Spring of Hangzhou is the best for brewing the West Lake Longjing Tea before they complement each other. This sweet and clear spring water can be presented to the guests for tasting.

- 置茶

用茶匙将茶叶从茶荷中拨入玻璃杯中，注意掌握茶与水的比例。置茶时心态平静，勿将茶叶落在杯外。

Tea Application

Push tea leaves from tea container to the glass cup with a tea spoon. Be aware of the tea-water ratio. In the process, keep a calm mindset and do not drop any tea leaf outside the glass.

- 润茶

向杯中注入1/4杯热水，使干茶吸水舒展。

Tea Moistening

Pour hot water into the glass to 1/4 full, let the dry tea absorb water and unfold.

- 高冲

高提水壶，让水直泻而下，接着用凤凰三点头的手法冲泡。凤凰三点头不仅是为了泡茶本身的需要，也是对客人表示敬意，是中国传统礼仪的体现。

Water Pouring from a High Position

Lift the teapot and let the water rush down into the glass. Then, brew the tea by using the technique called Phoenix Nodding Head Thrice. Such technique is for both meeting the need of tea brewing and showing respect to the guests. It represents traditional Chinese etiquette.

- 奉茶

  将自己精心泡制的清茶与宾客共赏。

  Presenting the Tea

  Present the carefully brewed tea to the guests.

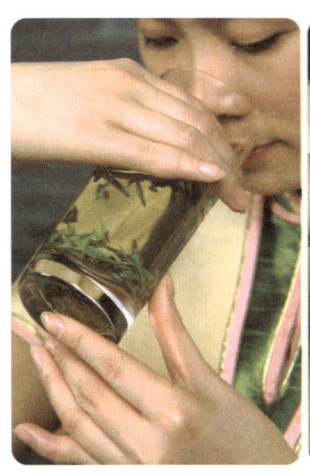

- 品茶

  龙井茶是茶中珍品，需细品慢啜，体会齿颊流芳、甘泽润喉的感觉。

  Tasting the Tea

  The Longjing Tea is a top-grade tea variety. Tasting slowly, you will feel the aroma flowing through your teeth and the sweetness passing by your throat.

> 冲泡白茶

冲泡白茶，茶叶的投放量与绿茶相当，每克茶的用水量约为50毫升。白茶的茶汁不易浸出，冲泡时间较长，适宜煎服，在中国浙江、福建一带生活的畲族有一套具有民俗特色的白茶茶艺。

> Brewing White Tea

In white tea brewing, the amount of the tea leaves applied is equivalent to that of the green tea. For each gram of tea, about 50ml water should be used. The taste of the white tea is harder to get loose and thus it requires a longer period of brewing. Cooking can be a proper choice. The She people living in Zhejiang and Fujian provinces keep a unique set of White Tea Art.

白茶茶艺——白毫银针茶艺

白毫银针是中国特有的茶品,它的特点是不炒不揉,白毫披身。

Tea Art for the White Tea—Tea Art for White-tipped Yinzhen Tea

The White-tipped Yinzhen is a tea peculiar to China. It is covered with white hair. Its making requires neither frying nor rolling.

- 备具

古朴的石台上放着石质茶海、陶质茶炉、陶质茶罐、陶质烤壶、陶质泡茶碗、陶质品茗杯、陶质奉茶盘、茶树枝搅茶棍、竹质分茶勺、竹制茶拨、铸铁煮水壶。

Tea Set Preparation

Put the entire tea set on a simple stone platform: stone tea tray, ceramic tea stove, ceramic tea can, ceramic baking pot, ceramic tea bowl, ceramic teacup, ceramic teaboard, a tea tree branch for stirring, bamboo tea spoon, bamboo tea stick, and cast-iron kettle.

- 烤壶

点燃茶炉,手执烤壶将其置于茶炉之上,以烘干烤热,受热宜均匀,烤炉温度不宜过高。

Baking the Pot

Light the tea stove, put the baking pot on the stove to bake it dry. The heating should be even and the baking temperature should not be too high.

- 投茶

壶烤热,将白毫银针拨入壶中。

Tea Application

Put the White-tipped Yinzhen Tea into the heated pot.

- 烤茶

用文火烘烤壶中之茶，使之香气焙出。

Baking the Tea

Bake the tea in the baking pot until the aroma comes forth.

- 闻香

轻闻茶香，判断干茶是否已经烤好。

Smelling the Aroma

Smell tea aroma and judge if the dry tea is properly baked.

- 入碗

将烤好的白毫银针投入碗中。

Into the Bowl

Put the baked White-tipped Yinzhen Tea into the bowl.

- 注水

将沸水悬壶高冲，注入陶质泡茶碗。

Water Pouring

Pour the boiling water from a high position into the ceramic tea bowl.

- 搅茶

用茶棍轻搅茶汤，令茶的味道厚重饱满。

Stirring the Tea Soup

Use the tea tree branch to stir the tea soup and make the tea taste thick and full.

- 分茶

用分茶勺将搅好的茶汤均匀地分到品茗杯中。

Tea Soup Allocation

Evenly allocate the stirred tea soup to the teacups.

- 敬茶

双手稳端奉茶盘，敬茶给客人。

Presenting the Tea

Firmly hold the teaboard with both hands and present the tea to the guests.

> 冲泡黄茶

冲泡黄茶时，投茶量与白茶相同，冲泡时水温以70℃为宜。

> Brewing Yellow Tea

In yellow tea brewing, the amount of the tea leaves applied is equivalent to that of the white tea. The best water temperature for brewing is 70℃.

黄茶茶艺——君山银针茶艺

君山银针是一种重在观赏的特种茶，其色、香、味、形俱佳，有"金镶玉"的美称。

Tea Art for the Yellow Tea—Tea Art for the Yinzhen Tea from Mt. Junshan

The Yinzhen Tea from Mt. Junshan is a special tea variety with a high value in appreciation. With high-quality color, aroma, taste, and appearance, it is hailed as Jade with Gold Inlay.

- 赏茶

将茶拨入茶荷以供欣赏，君山银针芽头挺直匀称，白毫完整鲜亮，色泽金黄。

Tea Appreciation

Put tea leaves in a lotus-shaped tea container for appreciation. The Yinzhen Tea from Mt. Junshan has straight and symmetrical buds, complete and bright white tips, and a golden color.

- 温杯

采用回旋斟水法温杯，提高杯身温度，清洁茶杯。

Warming up the Teacup

Warm up and clean the teacups by using rotary water pouring method.

• 投茶

将茶荷中的茶叶依次投入杯中，茶与水的比例约为1∶50。

Tea Application

Divide the tea leaves and put them into teacups. The tea-water ratio is about 1:50.

• 注水

先回旋斟水，再以悬壶高冲的手法注水至七分满。

Water Pouring

Pour water into the teacup in rotary and high-lifting methods and fill the teacup to seven tenths full.

• 静置

将茶汤静置片刻，可以欣赏到君山银针的茶舞。

Placed Still

Place the tea still for a while. The dancing tea leaves can be appreciated.

• 品茶

待茶芽大部分都立于杯底时便可观其色、闻其香、品其味了。

Tasting the Tea

When most tea buds stand on the bottom of the teacup, tea drinkers can observe tea color, smell tea aroma, and taste tea flavor.

> 冲泡乌龙茶

乌龙茶的冲泡是最为讲究、最为复杂的。一般冲泡乌龙茶，宜选用紫砂壶，同时根据品饮人数选用大小适宜的壶，因乌龙茶的品种较多，茶叶外形具有较大的差异，所以不同的乌龙茶的投放量不同。条形的半球形的乌龙茶，用量以壶的二三成满即可；松散的条索形乌龙茶，用量以壶的八成满为宜。乌龙茶茶艺又称"工夫茶"，而从某种意义上来说，工夫茶是现代茶艺的起源。

> Brewing Oolong Tea

Brewing of the Oolong Tea requires the best-planned and most complicated procedures. Normally, the purple sand teapot is the best for brewing the Oolong Tea. The size of the teapot is chosen based on the number of the drinkers. There are many Oolong Tea varieties and they vary widely in appearance. Different oolong teas have different application amounts. For the strip-shaped semisphere Oolong Tea, the tea leaves should fill 20% to 30% of the teapot. For the loose strip-shaped Oolong Tea, the tea leaves should fill 80% of the teapot. The tea art for the Oolong Tea is called Gongon Tea Art. In a sense, it is the origin of modern tea art.

乌龙茶茶艺——安溪铁观音茶艺

安溪铁观音是乌龙茶中的佼佼者。冲泡后汤色金黄明亮，香气馥郁持久，滋味醇爽甘鲜。

Tea Art for the Oolong Tea—Tea Art for Tieguanyin Tea from Anxi County

Tieguanyin Tea from Anxi County is the best of the oolong teas. Its tea soup is golden bright. Its aroma is strong and long-lasting. Its taste is mellow, brisk, sweet, and fresh.

- 温具

用热水冲烫茶具，一为清洁器皿，一为提高茶具温度。

Warming up the Tea Set

Pour hot water onto the Tea Set to clean it and warm it up.

- 赏茶

用茶匙取出适量茶叶置于茶荷中，让宾客观赏其外形。

Tea Appreciation

Take proper amount of tea leaves with a tea spoon and put them in a lotus-shaped tea container. Present the tea leaves to the guests for them to appreciate the appearance of the tea.

- 投茶

将茶荷中的茶叶拨入紫砂壶中。

Tea Application

Push the tea leaves into the purple sand teapot from the container.

- 注水

在壶满后继续注水，滤掉浮沫杂质。

Water Pouring

Continue to pour in water after filling so as to get rid of the foam and impurity.

- 浇壶

盖上壶盖，沿着茶壶外围再浇淋一些热水，以使茶壶里外温度一致。

Showering the Teapot

Put on the teapot lid, shower the tea pot with hot water to keep the temperature inside and outside the teapot even.

- 出汤

出汤的时间视茶老嫩程度及投茶量多少灵活掌握。出汤前，先将壶置于茶巾上沾干壶底水渍，然后再将茶汤倒入公道杯。

Tea Soup Harvesting

The moment for tea soup harvesting depends on the tenderness and quantity of the tea leaves brewed. Before harvesting, put the teapot on a towel to dry up the bottom and pour the tea soup into the Gongdao (Justice) Cup.

- 斟茶

将公道杯中的茶汤均匀分入闻香杯，再将品茗杯盖在闻香杯上翻转过来。

Tea Soup Pouring

Divide the tea soup evenly into the aroma-smelling cup and cover it with the drinking teacup and overturn them.

- 敬茶

将冲泡好的茶汤摆入茶盘，敬奉给客人。

Presenting the Tea

Put the filled tea cup on the tea tray and present it to the guest.

- 品茶

先拿起闻香杯闻取独特茶香，再端起品茗杯细细啜饮茶汤。

Tasting the Tea

Take up the aroma-smelling cup to smell the unique tea aroma. Then, take up the drinking teacup and taste the tea soup in small sips.

> 冲泡红茶

冲泡红茶有两种方法：清饮泡法和调饮泡法。清饮泡法每克茶用水量以50～60毫升为宜，如先用红碎茶则每克茶用水70～80毫升；调饮泡法是在茶汤中加入糖、牛奶、蜂蜜、柠檬等调料，茶叶的投放量可随品饮者的口味而定。

> **Brewing Black Tea**

There are two ways to brew the black tea: brewing pure tea and brewing laced tea. To brew pure tea, each gram of tea leaves needs 50ml to 60ml of water. If the broken black tea is brewed first, each gram of tea leaves needs 70ml to 80ml of water. To brew laced tea, the condiments such as sugar, milk, honey, and lemon juice are added into the tea soup. The quantity of the tea leaves depends on the taste of the drinker.

红茶茶艺——祁门红茶茶艺

祁门红茶具有香高、色艳、味醇的特点,冲泡时水温以90℃为宜,多采用白瓷杯冲泡。

Tea Art for the Black Tea—Tea Art for the Keemun Black Tea

The Keemun Black Tea features high aroma, bright color, and mellow taste. The best brewing temperature is 90℃. Normally, white porcelain teacup is used.

- 赏茶

祁门红茶条索紧秀,锋苗好,色泽乌黑润泽,干茶有灰色光泽,俗称"宝光"。

Tea Appreciation

The Keemun Black Tea has tight and slender strips and good tips. It looks black and moist. The dry tea has a gray luster, known as "Treasure Light".

- 投茶

用茶匙将红茶拨入壶中。

Tea Application

Push the black tea into the teapot with a tea spoon.

- 温具

用初沸之水,温杯烫盏,以提高壶的温度。

Warming up the Tea Set

Pour boiling water into the tea set to raise its temperature.

- 注水

用沸水悬壶高冲,令茶叶在水的激荡下充分浸润,利于色、香、味的充分发挥。

Water Pouring

Pour in boiling water from a lifted teapot, let the tea leaves fully tumble in the turbulent water, and fully set free the tea leaves' color, aroma, and taste.

- 温杯

用烫壶的水继续烫杯，以提高茶杯的温度。

Warming up the Teacup

Keep on warming up the teacup with the water having warmed the teapot.

- 分茶

将泡好的茶汤倒入公道杯中，再均匀地分入各个品茗杯中。

Tea Soup Allocation

Pour the brewed tea soup into the Gongdao (Justice) Cup and then allocate it evenly into the drinking teacups.

- 敬茶

将分好的茶敬奉给客人品饮。

Presenting the Tea

Present the tea to the guest.

- 品茶

祁门红茶香气甜润中蕴藏着花香，汤色红艳亮丽，杯沿有一道明显的"金圈"，品之滋味醇厚，回味绵长。

Tasting the Tea

The Keemun Black Tea has a sweet and moist aroma, which contains flower fragrance. The tea soup is brilliantly red and bright. Along the rim of the teacup is an obvious "Golden Ring". It tastes mellow with a lingering aftertaste.

> 冲泡黑茶

黑茶质地较为坚硬，为了使茶叶中的营养成分充分溶解，一般采用壶泡的方式，且宜用现沸的开水冲泡，茶叶的投放量以壶容量的三四成为好。另外黑茶散茶也可用盖碗品饮，一般投茶量为5~8克。

> Brewing Dark Tea

The dark tea has a hard texture. To fully dissolve the nutrients in the tea leaves, teapot brewing is usually adopted and the fresh boiling water is used. The tea leaves applied should occupy three or four tenths of the teapot volume. In addition, the bulk dark tea can be brewed with a tureen. In that case, five to eight grams of tea leaves should be applied.

黑茶茶艺——烤普洱茶茶艺

普洱茶具有耐泡性，可采用现烤现泡的方式冲泡。

Tea Art for the Dark Tea—Tea Art for Baked Pu'er Tea

The Pu'er Tea lasts long in brewing and can be brewed immediately after baking.

- 汲水

取水存于壶中待用。香茗宜用活水烹煮，更能展现普洱"陈韵"。

Taking Water

Take water with a teapot. Running water is the best for brewing the Pu'er Tea. It can better display the aged charm of the Pu'er Tea.

- 温壶

用文火将壶烤热。

Warming up the Teapot

Bake the teapot with soft fire.

- 赏茶

将散茶置于茶荷中，以供欣赏。

Tea Appreciation

Put the bulk tea in the lotus-shaped tea container for appreciation.

- 投茶

用茶匙将普洱茶拨入壶中。

Tea Application

Push the Pu'er Tea into the teapot with a tea spoon.

- 烤茶

将盛茶之壶放在烤炉上烘烤。

Baking the Tea

Put the teapot containing the tea leaves on the stove for baking.

- 注水

悬壶高冲，壶中注满水，用壶盖将浮在壶表面的泡沫刮去。

Water Pouring

Pour water from a high position to fill the teapot. Scrape away the foam on the surface with the teapot lid.

- 煎茶

将茶壶再次放在烤炉上烘烤。

Cooking the Tea

Baking the teapot again on the stove.

- 分茶

将壶中茶汤均匀倒入茶杯中。

Tea Soup Allocation

Pour the tea soup evenly into the teacups.

少数民族的茶艺
Tea Art of Ethnic Groups

白族三道茶

白族三道茶，是一种流行于云南省大理白族地区的民族茶文化，为白族待客的隆重礼节。三道茶分别为苦茶、甜茶和回味茶，白族人民将人生之道寄予茶中，认为人生应该"一苦、二甜、三回味"。

三道茶的制作很特别，且每道茶所用原料都不一样。第一道茶为"清苦之茶"，用的是云南大理产的感通茶，寓意人在青年时期要吃得起苦，勇于艰苦创业。第二道茶为"甜茶"，用的是下关产的沱茶。客人喝完第一道茶后，主人重新用小砂罐置茶、烤茶、煮茶，并在茶盅内放入适量的红糖、桂皮、核桃仁等。此道茶味道甜蜜清香，寓意人到中年以后就开始开花结果，有所收获了。第三道茶为"回味茶"，寓意人步入老年后，什么都要看淡些，回味人生之路是怎么走过来的。其煮茶方法与第二道相同，茶盅内还可放一些蜂蜜、碎乳扇片、花椒粒等辅料。饮这道茶时，一定要使茶汤和佐料均匀混合，故喝茶时可轻晃茶盅，最好趁热喝下。

白族人为客人敬茶是有讲究的，茶斟八分满，双手捧杯举至齐眉，恭敬地向客人道声"请"，才能将茶递到客人手中。

"Three-course of Tea" of the Bai People

The Three-course of Tea represents the tea culture in the area inhabited by Bai people in Dali, Yunnan Province. It is a solemn ceremony of the Bai people to receive guests. The Three-course of Tea contains the bitter tea, sweet tea, and aftertaste tea. The Bai people use the tea to symbolize the process of human life: bitterness of youth, sweetness of middle age, and aftertaste of old age.

The Three-course Tea is made through a unique process. Each course of tea uses different raw materials. The first course, bitter tea, is made

• 白族三道茶
"Three-course of Tea" of the Bai people

with the Gantong Tea produced in Dali Bai Autonomous Prefecture of Yunnan Province. This course means that young people should be brave enough to accept the bitterness of life and endure hardship during early development. The second course, sweet tea, is made with the bowl-shaped tea produced in Xiaguan. After the guest has finished the first course, the host puts the tea of the second course into a can, bake it, and cook it. Then, the host adds some brown sugar, cassia, and walnut seed into the tea. This course is sweet and fragrant, meaning the middle-aged people can have some harvest after bitter early years. The third course, aftertaste tea, symbolizes the old years of a person, who should be able to view everything lightly and review his early years from a more philosophical angle. This course is made in the same way as the second course. This time, some honey, broken milk fan, and pepper grain can be added into the tea and evenly mixed up with the tea soup. During drinking, the drinker can gently shake the teacup and drink the tea when it is still warm.

The Bai people keep certain protocol in presenting the tea to the guests. They fill the teacup eight tenths full, hold it above his eyebrow, and respectfully say to the guest: "Please enjoy the tea." After this, they hand over the teacup.

藏族酥油茶

藏族人民生活在风雪高原，以放牧和耕种旱地作物为主，常年食用以奶制品、牛羊肉和青稞为主的食物，很少食瓜果蔬菜。人之健康重在平衡，茶即是平衡藏民健康之物。所以藏族有"宁可三日无粮，不可一日无茶"之说。

藏族人民喝茶名目繁多，有盐茶、奶茶、酥油茶，其中以酥油茶最为普遍，有客人时常以酥油茶款待，以示主人的盛情。相传，酥油茶是文成公主创制的。她与藏王和亲时带到藏区大量茶叶。因为她喝不惯当地的牛羊奶，就将茶叶与奶混合熬制，后来就成了现在藏民爱喝的酥油茶。

酥油茶的制法很讲究，先将砖茶或沱茶捣碎，放到锅中熬煮，大约半小时后，即可将滤过的茶汁倒入特制的打茶筒内。打茶筒多为碗口粗、半人高的圆柱形。同

▶ **打酥油茶**
Making the Buttered Tea

时放入适量的酥油和佐料，佐料分很多种，可根据个人口味的不同添加，有食盐、糖、芝麻粉、花生仁、松子、鸡蛋等。然后趁热将一根搅棒伸入筒内，不停地搅动，目的是让酥油、佐料、茶充分融合在一起。酥油茶最好趁热喝。

  藏族人民用酥油茶待客很重礼节。待宾客坐好后，主人会将一个盒子放在桌子中间，盒子里装有用炒熟的青稞粉与茶汁捏成的粉团糌粑，同时摆好茶碗。主人倒酥油茶时，会按辈分的大小，先长后幼——倒上酥油茶，并热情地邀请客人用茶。然后主客一边喝酥油茶，一边吃糌粑交谈。

Buttered Tea of Tibetans

Tibetans reside in windy and snowy plateau and make a living by practicing livestock farming and planting dry-land crops. Their major food includes dairy products, beef and mutton, and highland barley. They seldom have melons, fruits, or vegetables. Tea is their secret to keep a healthy balance. Tibetan people have a saying: "It is better not to have food for three days than not to have tea for one day."

  Tibetan people drink various kinds of teas, including the salt tea, milky tea, and buttered tea. Of these, the buttered tea is the most popular. The host always treats the guest with buttered tea to show his hospitality. It is said that the buttered tea was invented by Princess Wencheng. When she got married to the king of Tubo, she brought with her large amount of tea. As she was not used to local cow and sheep milk, she mixed tea and milk together and made the buttered tea, which has since been loved by Tibetan people.

  The buttered tea is made through meticulous procedures. First, pound the brick-shaped tea or bowl-shaped tea into pieces and stew them in a cauldron for about half an hour. Then, pour the filtered tea soup into a specially-made tea-making tube, which is usually a half-man-tall cylinder having the diameter of a bowl. Meanwhile, put proper quantity of butter and condiments into the tube. There are various kinds of condiments, including salt, sugar, sesame powder, peanut, pine nut, and egg, and they can be added based on personal taste. While the soup is still hot, stir it with a stick. The nonstop stir is to fully blend the butter and condiments with the tea. The buttered tea tastes the best when it is still hot.

  Tibetan people have a set of protocol in treating guests with the buttered tea. After the guests are seated, the host will put a box in the middle of the table, which contains some tsamba kneaded with the fried highland barley powder and tea soup. The tea bowls are also ready for use. The host pours the buttered tea for the guests in order of their ages. The elder one gets the tea first. Then, the host warmly invites the guests to enjoy the tea. Tsamba is also offered. The host and the guests can talk happily while drinking and eating.

蒙古族奶茶

  蒙古族人民特别好饮茶，早晨起来即会煮一壶奶茶，一边喝着热腾腾的奶茶，一边吃些炒米、黄油、奶酪等，权当早餐了。不仅如此，他们中午和晚上还要各喝一次，有的甚至将茶锅置于炉上温热，以便随时饮用。

奶茶的制作分为普通奶茶和高级奶茶两种。普通奶茶制作较简单，将茶砖捣碎后放入锅中，加水煮沸，滤去茶渣。然后，往锅中倒入牛奶，茶和奶的比例为5:1，充分搅拌即可饮用；高级奶茶的制作较为繁琐，最好选用红砖茶，将其捣碎后用铜壶煮沸。滤出澄清的茶汁，将其倒入干净的桶中，上下搅动。然后，再将浓茶倒入锅中，加上一些调料煮沸即可饮用。调料可根据个人口味添加，有牛奶、羊奶、黄油、葡萄、蜂蜜、盐巴、肉类、谷物等。

蒙古人好客如好茶，不管认识不认识，只要有客人进入蒙古包内，主人就会奉上一碗飘香的奶茶。面对如此盛情的主人，客人应恭敬受之，才是对主人的尊重。

● 蒙古族奶茶
Milky Tea of Mongolian people

Milky Tea of Mongolian People
Mongolians like tea very much. They cook a pot of milky tea every morning. Their breakfast includes fried rice, butter, cheese and the milky tea. They also drink the milky tea at noon and in the evening. Some people even leave their teapot on the stove to keep it warm for drinking at any time.

There are two kinds of milky tea: ordinary one and first-class one. It is easy to make the ordinary milky tea: the brick-shaped tea is ground to pieces and put into a cauldron for cooking in water. After the tea dregs are filtered away, pour milk into the cauldron and make tea-milk ratio reach 5 to 1. After thorough stirring, the tea is ready for drinking. The first-class milky tea is made with the brick-shaped black tea through a more complicated procedure. First, the tea is ground to pieces for cooking in a copper teapot. Then the clear tea soup is filtered into a clean bucket and stirred up and down. Then, the strong tea is poured into a cauldron and cooked with some condiments to boiling. Then, it is ready for drinking. Condiments include cow or sheep milk, butter, grape, honey, salt, meat, and grain, and are added according to personal taste.

Mongolian people are hospitable. Anyone entering their yurt will be offered a bowl of fragrant milky tea; no matter whether they know this person or not. The guest should respectfully accept the tea; this is the best way to repay the host's hospitality.

傣族竹筒茶

竹筒茶是云南省西双版纳地区傣族别具一格的茶饮方式，同时也是傣族人待客的高规格礼节。

竹筒茶的制作以新鲜的香竹为最好，削去上节留下节，然后将鲜嫩的茶叶放入竹筒内，装满后，将竹筒放在火上烘烤，每隔几分钟要翻动一次，为的是受热均匀。待筒内茶叶慢慢萎缩，便将一根木棒插入筒内，用力压紧茶叶，再添进新茶。再烤再紧压再添茶叶，如此反复操作，直至竹筒内塞满茶叶。当竹筒外壁的颜色由绿变黄之时，说明茶叶烤得差不多了。取茶时要用刀砍破竹皮，用力要恰到好处，否则一个成型的圆柱茶就不完美了。这时的茶叶要马上冲泡味道才最好，既有茶的醇厚，还带有竹子的清香。

Bamboo Tube Tea of the Dai People

Bamboo tube tea is made by the Dai people living in Xishuangbanna Dai Autonomous Prefecture of Yunnan Province. Offering bamboo tube tea is a high-standard etiquette of the Dai people in receiving guests.

The fresh fragrant bamboo is the best for making bamboo tube tea. The upper bamboo section is cut off and the lower section is used as the bamboo tube. Then, insert the fresh tea leaves into the tube and bake it on a fire. Turn it around every several minutes to heat it evenly. After the tea leaves slowly shrink, insert a stick into the tube to compress them and add more new tea leaves. Repeat the process for several rounds until the bamboo tube is fully compacted. When the outer wall of the bamboo tube turns from green to yellow, it indicates that the baking process is almost complete. To get the tea, the bamboo tube should be halved with a knife. The cutting force should be right to get a perfect tea cylinder. The tea brewed at

• 傣族竹筒茶
Bamboo Tube Tea of the Dai People

that very moment has the best taste, which contains both the tea's mellowness and the bamboo's fragrance.

傈僳族雷响茶

傈僳族雷响茶流行于云南省怒江傈僳族人民居住的地区，是一种颇具民族特色的饮茶方式。用大瓦罐烧开水，小瓦罐烤饼茶，茶烤出香味后将开水注入煮约5分钟。然后滤去茶渣，加少许酥油及炒过碾碎的核桃仁、花生米、盐巴或者糖等。最后再将钻有小洞的鹅卵石用火烧后放入茶罐中，用来提高茶汤的温度和融化酥油。由于鹅卵石在容器内作响的声音好像雷鸣一样，所以得名"雷响茶"。响过以后马上用木杵上下搅动，使酥油充分融于茶汁，便可趁热饮用了。

Thundering Tea of the Lisu People

The Thundering Tea of the Lisu People is popular in Lisu area in Nujiang Prefecture of Yunnan Province. It has distinct ethnic features. First, a large earthen jar is used to boil the water and a small earthen jar is used to bake the caky tea. After the baked tea sends forth aroma, pour in boiling water and boil the tea for five minutes. Then, filter away tea dregs and add in some butter and fried and ground walnut seed, peanut, salt, or sugar. Finally, bake some holed cobblestones in fire and put them into the tea jar to raise the tea soup's temperature and melt the butter. As the cobblestones make thunder-like sound in the jar, the tea is called the Thundering Tea. After the thundering, stir the tea soup with a stick to make the butter fully melt into the tea. Then you can enjoy the tea before it gets cold.

● 傈僳族迎客的少女
A Lisu Girl Greeting the Guest

擂茶

擂茶一般以生茶叶、生米、生姜为主要原料，经过研磨配制后加水烹煮而成。擂茶在湖南、湖北、江西、福建、广西、四川、贵州等省区最为普遍。根据各地人们不同的口味和习惯，擂茶除了用"三生"的原料以外，还会另外添加一些其他的东西，如芝麻、花生、玉米、盐或者糖等。比如安化擂茶，它的原料就是花生、黄米、黄豆、芝麻、绿豆、南瓜子和茶叶，加少许生姜、胡椒和盐等，将所有的原料都炒熟后放入擂钵中捣碎，然后将捣碎的原料放入烧沸的水中，搅拌均匀，熬煮片刻便可。

Pounded Tea

The pounded tea is made by grinding and cooking the materials such as the raw tea leaves, raw rice, and raw ginger (known as "three raw materials"). It is popular in Hunan, Hubei, Jiangxi, Fujian, Guangxi, Sichuan, Guizhou and other provinces and regions. In addition to the three raw materials, other condiments can be added based on the tastes and habits of different drinkers. These condiments include sesame, peanut, corn, salt, and sugar. For example, the Anhua Pounded Tea is made of peanut, coarse rice, soya bean, sesame, mung bean, pumpkin seed, and tea leaves, with small amount of raw ginger, pepper, and salt. All these materials are fried, ground in a mortar, and put in boiling water. The soup is stirred evenly and cooked for a while. Then, the tea is ready.

• 擂茶
Pounded Tea